European Union Spatial Policy and Planning

R.H. Williams

Paul Chapman Publishing Ltd
144 Liverpool Road
London
N1 1LA

British Library Cataloguing in Publication Data

Williams, R.H. (Richard Hamilton)
European union spatial policy and planning
1. Regional planning – European Union countries 2. Space (Architecture) – European Union countries
I. Title
711.4'094

ISBN 1 85396 305 4

Typeset by Dorwyn Ltd, Rowlands Castle, Hants
Printed and bound by Athenaeum Press Ltd.,
Gateshead, Tyne & Wear

A B C D E F G H 9 8 7 6

R H Williams MRTPI is Senior Lecturer in Town and Country Planning and Associate Director of the Centre for Research in European Urban Environments at the University of Newcastle upon Tyne, where he has taught courses on EU spatial policies for over 12 years. He is the author of numerous articles, papers and other publications on EU policy and European comparative planning, and with Barry Wood has edited the series on Urban Land and Property Markets in France, Germany, Italy, Netherlands, Sweden and UK (UCL Press). He has played an active role in the Association of European Schools of Planning as a member of the Executive Committee 1987–92, and in ERASMUS student exchange programmes.

Contents

Preface

This is a book about an integrated Europe. It takes the perspective 'l'europe mon pays'. It is about the institutions, policy instruments and spatial planning at the scale of Europe as a whole and of that of the European Union.

For some years I have been teaching a course on European Union institutions and spatial policies to town and country planning students in the University of Newcastle upon Tyne at postgraduate and undergraduate levels, often alongside comparative teaching about the planning systems of other European countries. However, the latter is quite distinct from the subject-matter of this book, which addresses issues in planning that are quasi-domestic rather than foreign. I have for some time felt that there is a lack of textbook material directed specifically at the objectives of my courses for planning students on EU spatial policies and institutions. There are plenty of excellent publications on the politics and government of the EU, on the geography of regional development and regional policies, but none with the particular focus and range of concerns of those professionally involved in planning and of those studying to qualify as professional practitioners in spatial planning. The solution to this problem was clearly that I should write such a book myself.

The name of the professional field with which I am concerned can cause confusion. In Britain the term town and country planning and the professional title Town Planner are relatively well understood, but planning degree courses take a number of other titles, for example urban and regional planning, city planning, environmental planning. Problems of translation complicate the issue considerably in the European context. I therefore use the term spatial planning as a relatively neutral and inclusive term for all the various styles and concepts of planning found in the EU, and to encompass all spatial scales from the local to the whole of Europe.

EU policies and programmes already form a familiar and important part of the working environment for local planning authorities and professional planners, whether in the context of the Directive on the environmental assessment of projects or the need to identify and present infrastructure projects in such a way as to qualify for ERDF or ESF grants. This is reflected, for example, in the RTPI education guidelines. Following the 1992 single market programme and the Maastricht Treaty, the prospect of 'ever closer union' suggests that there will be many more ways in which the EU level of government will matter in the domestic practice in planning as in many other walks of life. The Commission should be and increasingly is recognised essentially as part of the structure of government and network of governmental contacts, in just the same way as national government departments are.

Learning to operate professionally in this context, to think European and to understand the implications of this policy context as well as of the continental spatial scale will present a major challenge to all planners, students and planning educators.

This book is intended to help meet this challenge. It is written for those who wish to, or need to know more about this new context and be able to interpret Commission initiatives with the same level of understanding that they can already bring to bear on national government policy towards planning.

Therefore it begins at the beginning, taking the reader through the basic institutional structures indicating what needs to be understood in the spatial policy context. It then outlines the scope and evolution of well established existing policies and identifies the components of what is now emerging as a coherent body of spatial planning and policy for spatial planning at the scale of Europe as a whole following the SEM and Treaty on European Union and in the run-up to the 1996 Inter-Governmental Conference.

In doing so it aims to meet twin objectives. The first is to provide a basic guide to the EU and textbook on its policies and institutions in the field of spatial planning. The second is to draw together, on the basis of the evidence available to date, indications of the future developments in EU spatial policy and the role this may play within the overall EU policy agenda for the late 1990s, dominated as it will be by the questions of monetary union and enlargement.

The common thread throughout is the range of interests and activities of planners. The scope of this book is defined in terms of the professional responsibility of spatial planners and of the curriculum of professional education in spatial planning, rather than in terms of the classical divisions of academic material or divisions of governmental responsibilities. I felt that the way to achieve this in a clear and convenient manner for the reader was to adopt a 4-part structure. Part I introduces the EU as a jurisdiction and planning subject. Part II traces the emergence of Europe-wide spatial policies and outlines the established policy sectors of spatial planning interest. Part III is concerned with the practical questions often asked by students interested in working professionally in European affairs, and Part IV concludes with thoughts on future prospects.

The ultimate goal is to assist readers to achieve the mental reorientation and develop the imagination necessary to operate effectively as a European planner, and thus to make a small contribution within this professional field to the creation of a Europe without frontiers.

Although I have tried not to allow my personal political stance on the issue of European integration and of Britain's place in the EU to intrude in the main text, I should conclude this preface by acknowledging my own political stance on the question. This can most easily be done by quoting a remark from an academic colleague at London University which I took as a compliment after I had given a visiting lecture on a theme similar to that of this book, who referred to my 'barely disguised Euro-federalist tendencies'.

R.H. WILLIAMS,
Newcastle upon Tyne.
January 1996

Acknowledgements

Many people have contributed to my knowledge and thinking over the years about EU affairs, whether consciously or not. Among them, there are a number of people to whom I owe a particular debt of gratitude.

In Belgium, Marios Camhis, Jean-Francois Drevet, Derek Martin (now with the Rijksplanologische Dienst, Den Haag), Horst Reichenbach and Gerhard Stahl from the European Commission; and Louis Albrechts from the Catholic University of Leuven. In Finland, Pekka Virtanen of Helsinki Technical University and Merja Kokkonen of the University of Joensuu. In Germany, Egbert Dransfeld, Hartmut Dieterich, Klaus Kunzmann, Winrich Voss and Michael Wegener from the University of Dortmund; Peter Schön and Brigitte Schabhüser from the Bundesforschungsanstalt für Landeskunde und Raumordnung, Bonn. In Netherlands, Andreas Faludi, Arnold van der Valk and Wil Zonneveld and students of the University of Amsterdam. In Portugal, Artur Rosa Pires of the University of Aveiro. In Sweden, Thomas Kalbro and Hans Mattsson of the Royal Institute of Technology, Stockholm.

Newcastle colleagues and students have over several years helped in many ways. Among former students, Ceri Doyle, formerly of Glasgow City Council and Hamilton District Council, now European Officer for South Lanarkshire District Council; and Katharine Ridley, then of Durham County Council and now head of the European office in the University of Newcastle upon Tyne; plus members of my option class in 1994–5 for willing participation in a consumer-testing project as the structure of the book was being planned, Heidi Antrobus, John Pilgrim, Jill Rae, Robert Thorley and Louise Tilsted. I want to thank Judith Eversley of the Royal Town Planning Institute and Anne Ramsay of the European Documentation Centre, University of Northumbria, for help with information requests, and Bret Fleming for preparation of graphics. I also want to record my thanks to Marianne Lagrange and colleagues at Paul Chapman for their encouragement and toleration.

Financial support at various stages of the book's preparation is gratefully acknowledged. This has come from a British Council Treaty of Windsor grant, the teaching staff mobility Action of the ERASMUS programme, the German Academic Exchange Service, Helsinki Technical University, which appointed me as Visiting Professor in 1993, and the Centre for Research in European Urban Environments, University of Newcastle upon Tyne.

From all of these people, and many others, I have benefited from advice, comments and criticism over the time that I have been studying European Union affairs, and especially over the gestation period of this book. I am sure that the end product is the better for this advice. However, responsibility for the final text is, of course, mine alone.

In conclusion, I must pay tribute to my wife, Bernadette, for her encouragement, and the rest of the family, Aidan, Helen and Laura, for their understanding and support during the long period over the summer when my main companion always seemed to be the word processor.

R.H. Williams

Glossary of abbreviations

A	Austria
AESOP	Association of European Schools of Planning
AETU	Asociación Española de Técnicos Urbanistas (Spain)
ANU	Associazione Nazionale degli Urbanisti (Italy)
B	Belgium
becu	billion ecus
BMBau	Bundesbauministerium (Germany)
BNP	Bond van Nederlandse Planologen (The Netherlands)
BNS	Bond van Nederlandse Stedebouwkundigen (The Netherlands)
BVS	Belgische Vereniging van Stedebouwkundigen (Belgium)
CAP	Common Agricultural Policy
CDAT	Comité Directeur pour Aménagement du Territoire (CoE)
CEE	central and eastern Europe
CEMAT	Conférence Européenne des Ministres pour Aménagement du Territoire (CoE)
CH	Switzerland
CI	Community Initiatives
CIS	Commonwealth of Independent States (former USSR)
CLRAE	Congress of Local and Regional Authorities of Europe (CoE)
CoE	Council of Europe
Comecon	Council for Mutual Economic Assistance
CoR	Committee of the Regions
Coreper	Council of Permanent Representatives
CPMR	Conference of Peripheral Maritime Regions
CSD	Committee of Spatial Development
CSF	Community Support Framework
CUB	Chambre des Urbanistes de Belgique (Belgium)
D	Deutschland (Germany)
DDR	German Democratic Republic
DG	Directorate-General (of the European Commission)
DK	Danmark (Denmark)
DoE	Department of the Environment (UK)
DOM	Départements Outre-Mer (France)

E	España (Spain)
EAAE	European Association for Architectural Education
EACS	European Association of Chartered Surveyors
EAGGF	European Agricultural Guidance and Guarantee Fund
EAUE	European Academy for the Urban Environment
EBRD	European Bank for Reconstruction and Development
EC	European Communities
ECB	European Central Bank
ECJ	European Court of Justice
ECOS	Eastern Europe City Co-operation Scheme
Ecosoc	Economic and Social Committee
ECSC	European Coal and Steel Community
ECTP	European Council of Town Planners
ecu	European currency unit
EEA	European Economic Area
EEB	European Environment Bureau
EEC	European Economic Community
EFLWC	European Foundation for Living and Working Conditions
EFTA	European Free Trade Area
EIB	European Investment Bank
EIF	European Investment Fund
EMI	European Monetary Institute
EMU	economic and monetary union
EP	European Parliament
ERDF	European Regional Development Fund
ERES	European Real Estate Society
ERM	exchange rate mechanism
ESDP	*European Spatial Development Perspective*
ESF	European Social Fund
EU	European Union
Euratom	European Atomic Energy Community
F	France
FAB	Foreningen af Byplanlaeggere (Denmark)
FIFG	Financial Instrument for Fisheries Guidance
GATT	General Agreement on Trade and Tariffs
GDP	gross domestic product
GDR	German Democratic Republic
GPA	Greek Planners Association
GR	Greece
I	Italy
ICLEI	International Council for Local Environmental Initiatives
IDO	Integrated Development Operations
IFHP	International Federation for Housing and Planning
IGC	Inter-Governmental Conference

IMP	Integrated Mediterranean Programmes
INTA	International New Towns Association
IPI	Irish Planning Institute
IRL	Ireland
ISoCaRP	International Society of City and Regional Planners
ITT	information technology and telecommunications
L	Luxembourg
LRA	Local and Regional Authority
LGIB	Local Government International Bureau (UK)
mecu	million ecus
MEP	Member of the European Parliament
MHAL	Maastricht–Heerlen–Aachen–Liège
MoE	Ministry of the Environment (Denmark)
NAFTA	North American Free Trade Association
NIS	newly independent states (former USSR)
NL	The Netherlands
Nordec	Organisation for Nordic Economic Co-operation
NPCI	National Programmes of Community Interest
NUTS	nomenclature of territorial units for statistics
OECD	Organisation for Economic Co-operation and Development
OEEC	Organisation for European Economic Co-operation
OJ	*Official Journal*
OP	Operational Programmes
ORA	Orientierungsrahmen (Germany)
P	Portugal
PHARE	Pologne et Hongoire Assistance pour la Réstructuration Économique
PIE	*Perspectives in Europe*
QMV	qualified majority voting
RETI	Association des Régions Européennes de Technologie Industrielle
RICS	Royal Institution of Chartered Surveyors (UK)
RPD	Rijksplanologische Dienst (The Netherlands)
RTPI	Royal Town Planning Institute (UK)
S	Sweden
SDEC	Schéma de développement de l'éspace communautaire
SEA	Single European Act
SEM	single European market
SEPLIS	Société Européenne des Professions Libérales Indépendantes et Sociales
SF	Finland
SFU	Société Française des Urbanistes (France)
SMEs	small and medium-sized enterprises
SPD	single programming document

SPU	Sociedade Portuguesa de Urbanistas (Portugal)
SRL	Vereinigung für Stadt- Regional- und Landesplanung (Germany)
TACIS	Technical Assistance for the Commonwealth of Independent States
TENs	trans-European networks
TEU	Treaty on European Union
TTWA	travel-to-work area
UK	United Kingdom
UNCED	United Nations Commission for Environment and Development
UNECE	United Nations Economic Commission for Europe
USSR	Union of Soviet Socialist Republics
VFB	Vlaamse Federatie voor Planologie (Belgium)
WEU	Western European Union

Copyright Acknowledgements

1. The permission of the copyright holders is gratefully acknowledged for the use of material for the following figures and illustrations:

Figure 3.4 © reproduced with permission from the Local Government International Bureau's European Information Service.
Figure 6.2 © Kunzmann.
Figure 7.4, 7.6, 7.8 © with permission from the Office for Official Publications of the European Communities.
Figure 8.1 © Celtic Interreg.
Illustration (p. 75) with permission from Sedgefield District Council.

PART I

Context, institutions and concepts

1

Introducing the EU as a jurisdiction and planning subject

This book is about European spatial policy and planning. This is a new subject, but not as new as many may suppose. There is a substantial body of European Union (EU) policy-making which has specific spatial objectives. This body of policy has developed over many years. The issue of spatial disparities has accompanied the development of the EU throughout its history, although it has not always been expressed explicitly or in the vocabulary of spatial policy. Not surprisingly, spatial policy development has followed from the successive enlargement of the EU. The biggest challenge is still to come: monetary union. If this is to be politically and technically acceptable, policy instruments will be necessary to address the spatial disparities that are anticipated or which may be accentuated as a result.

During the 1990s, we have seen a resurgence of interest in national and supranational spatial planning, and in the preparation of spatial planning studies of Europe as a whole. Partly, this is a response to economic integration and the single market, improvement of transport networks and new infrastructure such as the Channel Tunnel. It can also be interpreted as recognition of the importance this sector of policy will have if the ambitions for economic and monetary union and enlargement are to be achieved by the end of the century. As Scott (1993a: 70) points out, this was already recognised at the time of the Single European Act 1987.

The idea that it is possible and worth while to prepare a spatial plan for the whole of Europe may nevertheless seem rather strange and hard to grasp. By integrating the explanation of well established sectors of EU spatial policy with an account of the development of Europe-wide spatial policy, it is hoped that the reader will come to an appreciation of the scope, limitations and prospects for this scale of planning, and an understanding of the logic behind specific sectoral policies of a spatial nature.

The Treaty on European Union, familiarly known as the Maastricht Treaty, includes reference in the English text to town and country planning, the equivalent text in some other languages being an exact translation of the term used in this book: spatial planning (for a full listing, see Figure 4.1, p.58). Inclusion in the treaty means that the subject-matter of this book is now within the competence or legal authority of the EU, and the Commission is therefore empowered to make proposals for planning legislation. In fact, any such proposal would be subject to unanimity, not majority voting (see Chapter 3), so nothing much will occur in the near future, but the toe-in-the-door principle often applies whenever new policy sectors are introduced in this way (Williams, 1993a).

Nevertheless, the Maastricht Treaty and the European Union it created provides the backdrop to this book. As the Dutch Prime Minister said of the city, alluding not simply to its location in The Netherlands: 'Maastricht is the hand which The Netherlands extends towards Europe.'

Major steps towards European Union have taken place following negotiations between the governments concerned, often at intergovernmental conferences (IGCs). These have been held in Messina in 1955, Luxembourg in 1985 and Maastricht in 1991 (see Chapters 2 and 3). It is widely expected that at the forthcoming IGC, due to start in 1996, further moves will take place to strengthen the legal powers giving the EU competence over spatial planning.

There has for many years been no shortage of EU policy coming within the scope of spatial planning as defined here (see below), for example in the fields of regional and environment policy and in the operation of the structural funds. Meanwhile, new powers and policy sectors created elsewhere in the Maastricht Treaty, such as the title on trans-European networks (TENs) (see Chapter 9), are likely to be of greater significance for planning than the insertion in the treaty of town and country planning will be. These themes, then, form the subject-matter of this book; it is not a critique of their operation and achievements. The main aim is to explain the scope and rationale of the whole package of spatial policy, convey an understanding of who is responsible for what, trace the emergence of EU spatial planning and make suggestions concerning its future importance.

Place of spatial planning in EU policy-making

Even the most enthusiastic planners may hesitate to claim that planning policy is as important a part of the European political agenda as foreign and taxation policy, as John Major seemed to imply in a campaign speech during the 1994 European Parliament elections. He claimed that anyone who does not vote in the European election would be allowing Britain to 'lose her freedom to decide her own taxation policy, *planning policy*, foreign policy and will have committed Britain to a single currency' (quoted in the *Guardian*, 23 May 1994, emphasis added). Nevertheless, it is an indication of its increasingly high profile in EU affairs that he should have drawn attention to it in this manner. An association between spatial policy and economic and monetary union (EMU) may not be so wide of the mark, a point returned to in the concluding discussion in Chapter 15.

In 1992, the then British Minister of State for Planning, Tony Baldry, saw it similarly in a speech to the Royal Institution of Chartered Surveyors, although it is not certain whether the implied criticism is directed at the Commission in Brussels or at the UK Department of the Environment in London's Marsham

Street: 'EC involvement in planning and development matters has been likened to an incoming tide. That tide shows no sign of ebbing. Instead we have an ever increasing flow of planning initiatives issuing from Brussels. Almost as many as from Marsham Street' (DoE press release, 1992).

The sentiment expressed by Baldry is likely to become increasingly true in the second half of the decade. In addition to the widely expected pressure from some member-state delegations to incorporate stronger powers over spatial planning in the EU treaties at the 1996 IGC, there is also substantial pressure to include an urban title in the treaty, or at least to incorporate explicitly a competence (i.e. legal powers) in urban policy and/or urban planning (see Chapter 11) in order to create the legal basis for an EU policy to address urban problems and local development.

Steps have already been taken to prepare for an increased role for the EU in spatial planning. The Commission published the *Green Book on the Urban Environment* in 1990 (Commission, 1990b; see Chapters 10 and 11), and the *Europe 2000* study, *Outlook for the Development of the Community's Territory*, in 1991 followed by the *Europe 2000+* study, *Co-operation for European Territorial Development* (Commission, 1991; 1994a; see Chapters 6 and 12).

Meanwhile, all member-states have senior officials sitting on the Committee of Spatial Development, which is preparing the *European Spatial Development Perspective* (see Chapter 12). A Co-operation Network of Spatial Research Institutes is proposed to be set up in 1996 (Chapter 3), and a *Compendium of Spatial Planning Systems and Policies* is in course of preparation for the Commission by consultants. The compendium for which the Commission's Regional Policy Directorate (DG XVI) let a contract in 1993 is intended to be a comprehensive source of information on the rules, regulations, procedures, policies and operation in practice of the land-use planning systems of all the member-states (see Chapter 12).

It should be clear, therefore, that EU spatial planning is a subject which is developing rapidly. There should be no doubt about the importance of the subject to anyone intending to be professionally active in spatial policy, planning and development.

The corollary is that any textbook such as this is confronted with the moving-target problem. As far as possible, this book is up to date to summer 1995, but readers should be aware that the theme of spatial policy and planning is one that is developing all the time and that factual and legal points can rapidly become out of date.

This book therefore aims both to guide the reader through the well established body of spatial policy and to convey an understanding of the scope, purpose and extent of spatial policy for Europe as a whole. It is hoped that, on this basis, readers will be able to understand and interpret for themselves future developments in EU spatial policy, and perhaps themselves be responsible for formulating them. Fundamentally, the educational purpose of this book is to encourage readers to think European in their approach to planning and spatial policy in all its many aspects: to think European about space and to think spatially about Europe.

Which Europe? Terminology

For the most part, this book concentrates on the EU, although several of the themes and policies apply beyond its borders, especially where the EU has concluded agreements with European Free Trade Area (EFTA) countries or prospective new member-states. The different groupings, and the 'new European architecture' which they now form, are discussed in Chapter 2.

The term European Union has only been the correct term since ratification of the Treaty on European Union in 1993. Strictly speaking, the term European Union refers to the three 'pillars' of the union, the European Community, Common Foreign and Security Policy, and Co-operation on Home Affairs and Justice (see Chapter 2). The subject-matter of this book falls within the first of these, the European Community.

For simplicity and to avoid confusion, however, the term EU is used throughout the book to refer to the European Union in whatever form it existed at the particular period under discussion, even though the terms European Economic Community (EEC) or European Communities (EC) might pedantically be correct. The only exception to this is when it is necessary to be precise in order to set out the chronology of its institutional development (chiefly, this applies to Chapters 2 and 3).

Thinking European

Much of the material in this book aims to convey an understanding of EU policies and programmes that are spatial in form, or which interact with local spatial policy because these are the themes which form the basis of everyday professional practice. A necessary foundation of this is the capacity to think spatially about the territory of the EU as a whole, and to interpret the policy formulation that is being undertaken at this spatial scale.

Imaginative thinking about the spatial structure and relationships within Europe may require some mental overturning of conventional geography. In terms of Europe's spatial structure, Britain is no longer an island. The Republic of Ireland sometimes regards itself as the only island member-state, and this is in the strict sense true if one accepts the Channel Tunnel as having ended Britain's island condition. However, for many EU purposes, Greece and Finland must also be regarded effectively as island states. In another sense, Switzerland is also an island, being the only west European country in neither the EU nor the European Economic Area (EEA), but totally surrounded by member-states.

At the EU scale, the spatial structure of the territory which forms the planning subject is itself subject to change to a much greater degree than is normally the case within national territories. This is partly the consequence of enlargement, partly of new infrastructure such as the Channel Tunnel, and partly of political decisions concerning the location of functions of European significance, as Chapter 6 shows.

Thinking spatially about Europe as a whole has been aided by the generation of spatial metaphors. The classic example is the Blue Banana (see Figure 1.1), a metaphor of French invention expressing the economic and political core of the EU and its periphery (Brunet, 1989). Other metaphors include the golden triangle and a bunch of grapes (see Chapter 6).

If all this seems strange, it is hoped that all will become clearer. First of all, it is necessary to establish what spatial policy and planning is taken to mean, what the scope of the subject is in the context of EU policy-making, how the book is organised and what it is trying to achieve.

Spatial policy and planning

A broad definition of the scope of spatial policy and planning is taken. The word spatial is used to express a focus on the location and distribution of activity within the territory or space of Europe. Spatial policy includes any policy designed to influence locational and land-use decisions, or the distribution of activities, at any spatial scale from that of local land-use planning to the regional, national and supranational scales.

Here, spatial policy is taken to include any EU policy which is spatially specific or is in effect spatial in practice, whether or not it is deliberately designed to be, and any policy which is designed to influence land-use planning decisions, to be integrated with local planning strategies or to be implemented by local and regional authorities of member-states as part of their spatial planning responsibilities. Spatial planning is more specifically defined as a method or procedure to influence future allocations of activities to space or space to activities. It makes use of urban and regional planning instruments to set out and implement spatial policy at whatever spatial scale.

Spatial policy and planning is not coterminous with regional policy and planning. Whereas the former embraces any spatial scale, regional is used here in the sense of policy applied to the scale of a region or regional authority of a country.

In addition to the formulation of Europe-wide spatial strategies, the scope of EU spatial policy as considered here includes funding programmes such as the European Regional Development Fund (ERDF) and the other structural funds, the Cohesion Fund and other Community instruments with specific spatial targets, TENs and communication infrastructure, crossborder planning initiatives, regulatory measures affecting land use, such as those adopted as part of the EU's environment policy, and networking and lobbying by authorities and interests concerned with spatial matters. Local and regional authorities are in effect the executive arm of the Commission, responsible for implementing several of these programmes.

One omission should be noted. Although the EU's Common Agriculture Policy (CAP) is of enormous importance and is spatial in its outcome, it is not discussed at length separately here. Instead, it is discussed as appropriate within the context of other policy sectors. Agriculture does not always fall within the above definition of spatial planning, being outside the legal scope of spatial planning instruments. More pragmatically, the majority of Europe's planning professionals work in urban and regional planning, and some selection is necessary in order to keep this book to a manageable length.

The subject-matter of this book is quite different from those which seek to explain the different spatial planning systems of the different member-states or undertake a crossnational comparative analysis of spatial planning law and policy in the various national systems. It does however draw upon the increasingly substantial literature on this type of study (see, for example, Williams, 1984; Davies, 1989; Wood and Williams, 1992; Dransfeld and Voss, 1993; Dieterich *et al.*, 1993/6) in order to identify issues which may become the subject of proposals for European legislation. The forthcoming *Compendium of Spatial Planning Systems and Policies* will massively increase the source material available for this type of analysis, as well as providing the Commission with the basis on which to develop proposals for legislation (see Chapter 12).

Differences between the national systems of spatial planning and of property development must be regarded as non-tariff barriers, which may need to be overcome if they represent impediments to, or distortions of, the economic integration of the EU (see Chapter 2). Questions arise, therefore, of the extent to which there is some convergence between planning systems in the different member-states (Healey and Williams, 1993) and whether any harmonisation measures should be introduced into planning systems in order to remove impediments to the single market. These and other issues for the future are discussed in the final chapter.

A Euroneutral rather than a British standpoint is taken throughout this book. As far as possible, a position of neutrality is adopted, considering Europe and/or the EU as a whole. Although some anglocentric bias is difficult to avoid, and occasionally introduced deliberately, especially in Part III, this book does not set out to examine the UK *vis-à-vis* the EU. There are a number of valuable publications which do this (see, for example, Haigh, 1989; RTPI, 1994).

Structure and logic of the book

The material in this book is organised into four parts. Part I introduces the EU as a jurisdiction and a planning subject. Part II focuses on the EU as one spatial planning subject and the emergence of a Europe-wide spatial policy that has occurred during the 1990s, outlining the spatial policy development and evolution of established policy sectors of spatial planning importance that has taken place over the last 20–5 years. Part III is concerned with practical questions of professional practice that students often ask concerning possibilities of working professionally on EU affairs or in another EU member-state. More specific guidance on information sources and study-abroad programmes is contained in the appendices. Finally, Part IV looks to the 1996 IGC and future prospects for EU spatial policy.

Within Part I, this chapter sets the scene. Chapter 2 discusses the nature of the EU, why it was created and how it differs from the many other supranational organisations in Europe and elsewhere with which it is sometimes compared. Some of the other supranational bodies in Europe, such as the Council of Europe, are of importance in relation to European spatial policy, and it is not unusual for local planners to be involved in Council of Europe activities from time to time, so their role is outlined as well. Chapter 2 also outlines the evolution, varying titles and successive enlargement of the EU, setting this in the context of present-day Europe and what is sometimes referred to as the new European architecture. The subject of EU spatial policy is linked to the changed structure of Europe and to theories of European integration, the creation of the single market and key words such as harmonisation and subsidiarity.

Chapter 3 outlines the role of the various institutions of the EU, in order to ensure that readers have a clear idea of which institution has what powers in relation to the spatial policy issued covered here, how European law is made and by whom, and the forms it may take. The importance of the treaties and their review at the 1996 IGC is also discussed. Reference was made above to the concept of considering national systems of spatial planning law and policy as non-tariff barriers. Chapter 4 draws attention to what seems to some people to be the greatest non-tariff barrier of all – language – whose significance paradoxically is often underappreciated by those coming from a monolingual background, as many English-speaking planners do.

The policy focus of Part II is introduced in Chapter 5 with an analysis of the development of EU sectoral and spatial policy, focusing on the chronology of those sectoral policies which have substantial sectoral components, and of the steps in the development of an EU spatial policy framework, culminating in the production of the *Europe 2000* and *Europe 2000+* reports (Commission, 1991; 1994a). This is related to the overall chronology of the EU and other supranational bodies found in Chapter 2. Chapter 6 seeks to promote the capacity to think continental in the context of Europe's spatial structure and policy on a more conceptual basis. The underlying theme is the learning process that planning as a whole has gone through in order to grasp the idea of supranational spatial policy for Europe.

For many, involvement in EU policy is motivated by the pot of gold at the end of the rainbow in the form of the funding programmes. These are the subject-matter of Chapter 7, especially the Cohesion Fund and the three structural funds: the European Regional Development Fund (ERDF), the European Social Fund (ESF) and the guidance section of the European Agricultural Guidance and Guarantee Fund (EAGGF).

Regional policy as understood by the Commission is more than just the structural funding programmes. There is a whole range of networking and crossborder planning initiatives. Networking of all sorts between municipal and regional authorities for policy and lobbying purposes has grown massively in the 1990s. Non-tangible forms of networking, lobbying and crossborder planning fall within the scope of Chapter 8, while physical network-building is treated in Chapter 9, which is about transport policy, infrastructure networks and the programme for TENs.

Chapter 10 concerns environment policy, the sector within which spatial planning, or 'town and country planning' in the words of the English-language text of the treaty, has for some time been associated, along with policy for the urban environment. There is now an enormous body of EU environment legislation, for which guidelines only can be provided here. In contrast, the subject of Chapter 11, EU urban policy, does not yet exist in a formal sense although many elements of such a policy are identified, and it may exist in a more formal sense after the IGC.

Part II concludes with a review in Chapter 12 of the state that EU spatial policy development has reached by the mid-1990s, prior to the IGC process itself.

Part III aims to address questions asked by students who wish to work professionally on EU affairs. Chapter 13 is concerned with another form of non-tariff barrier, professional titles, mobility and mutual recognition of professional qualifications, while Chapter 14 discusses the European liaison work of local and regional authorities.

Finally, in Part IV, Chapter 15 offers some suggestions on how EU spatial policy may develop during the IGC and beyond. It points out possible developments in spatial policy during the later 1990s that the reader who wants to keep up to date should look out for. Readers are warned that Chapter 15 is the chapter which will either prove prophetic or be rendered irrelevant by events.

Language and Euroenglish

It may seem to be stating the obvious to draw attention to the fact that this book is written in English. This is done in order to emphasise that language is a filter to understanding, especially whenever the subject-matter is not from the same country or language area, and to point out that much of the vocabulary that is used is in fact Euroenglish, a concept defined in Chapter 4.

I have used Euroenglish in the title: spatial planning. This could be taken to mean the same as town and country planning, and to refer to the functions of UK local planning authorities under the town and country planning Acts. It includes this, of course, but it means more than this. It means planning the spatial or territorial location of activities and physical development at any or all spatial scales, from that of the building precinct (for which formal and legally binding plans are required under some member-states' spatial planning systems, for instance the German *Bebauungsplan* and the Dutch *bestemmingsplan*), to that of regions, nations and Europe as a whole. Spatial planning is in fact an exact translation of the German *Raumplanung* and comes close to the sense of the French *aménagement du territoire*.

So what is Euroenglish? The term is defined here as the use of English words to convey non-British ideas and concepts. Euroenglish is used constantly in the context of EU affairs, often subconciously, and is an essential medium for the discussion of planning in a pan-European framework.

The use of Euroenglish is also a means of achieving the Euroneutral position this book seeks to adopt. If purely British terminology were to be used instead, a central teaching purpose of this book could be undermined. It would be as bad as using sexist masculine language when gender neutrality is required.

The professional education curriculum

Although this book is not intended to address a specifically British agenda, the British model of planning education is an important point of reference for those involved in curriculum development in several other European countries. For this reason, it is worth noting that the guidelines issued by the British Royal Town Planning Institute (RTPI) for the contents of the curriculum for accredited courses include 'The planning system in context . . . the UK in the EC [*sic*]; EC spatial policy; EC planning and environmental policy and legislation . . . comparison with other countries, especially the EC; with relation to environmental, urban, regional, rural and land use/transport planning matters' (RTPI, 1991: 3).

There are people professionally engaged in spatial planning in all EU member-states, but their educational background and the form and status of their professional titles vary greatly. One of the best accounts of the variety that exists in planning education in Europe is still to be found in Rodriguez-Bachiller (1988). The Association of European Schools of Planning (AESOP) has produced directories of planning education programmes offered by its members throughout Europe, which also demonstrates the great variety of educational concepts and provision of courses. AESOP has also put forward suggestions for curriculum development (Albrechts *et al.*, 1990) in order to meet the need for a European component in planning-education degree programmes such as that set out by the RTPI.

The European planning community

There is a growing sense in which it is possible to identify a European planning community as a network of people who are in contact with each other through either formal associations or through informal and often intangible links. One way this occurs is through the European liaison functions of local planning authorities and firms, participation in Council of Europe activities or EU lobbying and networking programmes such as RECITE or Ouverture, or through taking an active role in associations such as RETI or Eurocities (see Chapters 8 and 14).

Another manner in which a European planning community has developed is through the development of professional and academic bodies such as the European Council of Town Planners (ECTP) and AESOP, together with others that include aspects of planning within their scope (see Chapter 13).

National associations and institutes of spatial planning from any EU member-state can join the ECTP (see Chapter 13). The RTPI is the recognised professional body within the UK for planners, and is by far the largest member of the ECTP. The ECTP has adopted a charter defining the field and nature of the planner's profession and its education and professional conduct requirements. The membership of AESOP consists of over 160 university departments teaching spatial planning in over 30 European countries and regularly attracts 200–300 participants to its annual congresses.

The European planning community is an exciting and open environment in which to work. This book will have achieved its purpose if some readers join this community.

2

The quest for European union

After the end of the Second World War, there was very widespread determination that European conflict should never again be allowed to lead to such devastation. Political pressures to create new international institutions that would be effective in preventing renewed conflict were very powerful. Within Europe, it was recognised that conflict between European nations had, twice in the twentieth century, led to world war. The fundamental motive of politicians such as Jean Monnet, Robert Schuman and Paul-Henri Spaak whose drive and determination laid the foundations of the EU was to ensure that such conflict could never again happen.

This was made clear in the speech made by Robert Schuman, French Foreign Minister, at the Quai d'Orsay on 9 May 1950, which became known as the Schuman Declaration:

> the French Government proposes to take action immediately on one limited but decisive point . . . to place Franco-German production of coal and steel under a common High Authority, within the framework of an organisation open to the participation of the other countries of Europe . . . The solidarity in production thus established will make it plain that any war between France and Germany becomes not merely unthinkable, but materially impossible.

Although talking about the seemingly mundane and technical subject of coal and steel production, he went on to make radical proposals with far-reaching implications for national sovereignty and the whole future of European integration: 'By the pooling of basic production and the establishment of a new High Authority whose decisions will be binding on France, Germany and the countries that join them, this proposal will build the first concrete foundation of a European Federation which is indispensable to the preservation of peace' (Monnet, 1978: 298; Holland, 1994: 25).

The EU is not the only outcome of this political drive. Many other supranational associations of European states were proposed or established. Nor was there a direct evolution from the early ideas via the Treaty of Rome to the present union. There were several different ideas about how European integration should be achieved, and different ideas concerning the appropriate institutional structures and degrees of economic and political integration. Several approaches were tried. Some proved to be false starts, such as the European Defence Community. Others evolved into the other European supranational institutions we have today, such as the Council of Europe (CoE) and the Organisation for Economic Co-operation and Development (OECD).

It cannot be emphasised too strongly that the EU is qualitatively different from all of these, for reasons set out in the second extract from the Schuman declaration quoted above and discussed in the context of the theory of integration, below. Creation of the European Coal and Steel Community (ECSC) was the single decisive act referred to in the Schuman Declaration, and formed the foundation on which the EU that now exists was built. Just as decisions of the proposed High Authority of the ECSC were to be binding on the individual countries, so the EU is a jurisdiction with legally binding powers and therefore has a quasi-domestic governmental role. Its significance to spatial planners ultimately is the consequence of its status as a jurisdiction.

In contrast to the EU, all the other supranational bodies discussed here are intergovernmental in their operations. Member governments therefore are free to participate or not in their various activities without calling into question the whole philosophy of that body.

The objective of this chapter is to provide a picture of the different ways in which the map of Europe is now divided up, a guide to who does what in relation to spatial planning at the supranational scale, an outline of the stages in the development of the EU from the original concept at the 1955 Messina Conference to the present union of 15 member states, and an understanding of why the EU is different from all the other European supranational organisations.

This chapter provides a brief outline of these other organisations, dwelling in more detail on those which have played a role in spatial planning. These are categorised into those originating in the years following the Second World War, and those which are contemporary with, or the consequence of, the collapse of communism in 1989–90. Within each category, a loosely chronological basis is adopted. It then sets out the distinctive features of the EU which make it qualitatively different from the other bodies described here. Brief reference is also made to the North American Free Trade Association (NAFTA) because there is some evidence already that NAFTA may cause north American policy-makers to think more in spatial policy terms than they have hitherto. The EU may therefore be a model for NAFTA.

While this is not a political history or theory text, it does aim to provide the reader with an understanding of the nature of the political institutions, whether of the EU or of other supranational bodies, to which specific initiatives and spatial policy developments can be related. In order to reinforce the distinction between the EU and other intergovernmental bodies, it is necessary to start with a little theory.

Theory of European integration

There is a body of political theory which has been developed as the conceptual foundation for European integration and as an attempt to explain the political and institutional processes involved. The underlying logic or rationale for many specific measures derives from this body of theory. This applies to such high-profile issues as the possible creation of a single European currency, but also to matters of more immediate concern to planners as it may be argued that some degree of harmonisation of planning procedures follows from the logic of European integration.

In essence, the postwar drive to bind the nations of Europe together and remove the threat of another European war manifested itself in two fundamentally different approaches to European integration, the confederalist and the federalist (Borchardt, 1995). The confederalist approach obtained whenever individual countries agreed to co-operate without ceding sovereignty. Co-operation therefore was intergovernmental in its nature. The CoE and the OECD, discussed below, are examples. The confederalist model also applies to the two pillars of the EU which are outside the scope of this book, the second on Common Foreign and Security Policy, and the third, Co-operation in Justice and Home Affairs.

The federalist approach aims to dissolve the traditional differences between nation-states, overcoming the dangers of domination by one over another which has too often been a cause of European conflict. This is done by pooling sovereignty, giving political responsibility to common institutions. A necessary compromise from the outset was that there should not be an immediate abandonment of statehood, but that transfer of power to common institutions should take place gradually, with an emphasis on those powers necessary to achieve agreed common objectives. Thus, participant nations were not required to relinquish sovereignty in its totality. Instead, they were required to abandon the doctrine of its indivisibility.

The practical effect of this is felt in the impact of EU legislation. Whereas resolutions of confederal or intergovernmental bodies are legally binding, if at all, to the governments of member-states to whom they are addressed, the doctrine of 'direct effect' applies to EU law. In other words, it applies directly to individual citizens and legal bodies within member-states. Local and regional authorities are responsible for implementing many of the policy instruments in the spatial sectors covered here. In a sense, therefore, local and regional authorities act as the executive arm of the EU government or Commission, just as they may for their national government. Another word for this is subsidiarity, the post-Maastricht key word.

Tension between pursuing a grand design and step-by-step incrementalism emerges in successive debates on European integration. One approach was to identify the features that bound together existing large federations such as the USA, ranging from a common currency and foreign policy to open internal borders and common standards and procedures. Another was the functionalist approach of examining which sectors of the economy could benefit from closer integration, and to create the institutional structures to administer this. Monnet was in many ways the most influential of the founders of the EU, his methods containing elements of both approaches. Holland (1994) reviews different theories of integration, arguing that the ideas developed in the 1950s and 1960s have not stood the test of time and events, and suggesting that a neofunctionalist theory should be developed to meet the need for a conceptual framework to order understanding of the processes at work. Pinder (1991) explores the meaning of federalism in the EU context and puts forward seven propositions synthesising the federalist and neofunctionalist ideas as the neo-federalist idea.

Postwar institution building

The Benelux Agreement

The agreement to form the Benelux Union, a customs union among Belgium, The Netherlands and Luxembourg, must be mentioned first both for chronological reasons and because of its importance as a model for later developments. In the context of spatial policy, it is still providing a test-bed for the development of transnational planning concepts in the 1990s (see Chapter 6).

In fact, it is rather misleading to classify the Benelux Union as a postwar creation. Its precursor was the agreement of the Belgian and Luxembourg governments in 1921 to remove trade restrictions between each other. The respective governments of all three Benelux countries, in exile in London, negotiated an agreement in 1944 to create a customs union in which internal tariffs would be greatly reduced and a common trading position presented to the outside world.

Their ambitions did not whither in the face of the reality of government after the war. A common commercial policy was adopted in 1950, liberalisation of capital movements in 1954 and free movement of labour in 1956. Throughout the history of the EU, the level of integration of the Benelux countries has been higher than that of the EU as a whole. This has not had a detrimental effect. On the contrary, the Benelux countries have performed the role of models for other member-states to follow, and of guardians of the ideals of Monnet and Schuman (Holland, 1994: 24).

Council of Europe

The CoE was the first of the postwar supranational bodies to be established. It came into existence on 3 August 1949 with the objectives of promoting greater unity and co-operation between the peoples and nations of Europe. Membership is open to any European country that has a democratic form of government, respects the rule of law and protects human rights. The original signatories were Belgium, Denmark, France, Ireland, Italy, Luxembourg, The Netherlands, Norway, Sweden and the UK. By 1995 there are 34 members. Nine former Communist Party states are already among these and another six are seeking membership. It operates in two official languages, English and French, and is based in Strasbourg, where it plays hosts to the monthly plenary sessions of the European Parliament in its Council Chamber, an arrangement the French government and the city of Strasbourg go to great lengths to maintain in spite of many pressures to consolidate the location of EU institutions in Brussels.

At the time of its foundation, some members saw it as the body through which economic and political integration of Europe may be achieved, while others saw it as an association of independent states who co-operated as far as they saw fit on a range of cultural, scientific and human rights issues. With the foundation of the ECSC and EEC, the former view was overtaken by events and the CoE has followed the latter course, becoming intergovernmental in its mode of operation.

The CoE is sometimes dismissed from serious consideration because of its lack of formal powers. It is sometimes seen as the promoter of town twinning, school exchanges and councillors' junkets. It has, however, undertaken many activities in the spatial planning field and has many achievements to its credit. Apart from the EU itself, planners working for local and regional authorities are more likely to be involved in CoE initiatives than those of any other supranational body.

The CoE has three main institutions, the Committee of Ministers, the Parliamentary Assembly and the Congress of Local and Regional Authorities of Europe. The Parliamentary Assembly, consisting of delegations appointed by the national parliaments of CoE members, is the CoE's main policy-making body. It has 13 specialist committees, one of which is concerned with regional planning and local authorities, the Comité Directeur pour l'Aménagement du Territoire (CDAT). The assembly makes recommendations to the Committee of Ministers. The latter is the decision-making body and is composed of the foreign ministers of the members.

Decisions by the committee may take the form of recommendations to member-states or conventions which are binding only on those member-states that chose to sign them. One of the most notable of these is the European Convention of Human Rights, signed in 1950, under which the CoE has set up the European Court of Human Rights and the European Commission of Human Rights. Judgments of the court are binding on all parties. It has happened that cases involving disputes over local planning issues have reached this court, which is also located in Strasbourg. It should not be confused with the European Court of Justice in Luxembourg, which is an EU institution (see Chapter 3).

Within the framework of the CoE, several standing conferences of ministers with other portfolios relating to its activities have been created, among them the Conference of European Ministers of Transport and the Conference of European Ministers of Spatial Planning. Like many CoE bodies, this is known by its French acronym as CEMAT (Conférence Européenne des Ministres pour Aménagement du Territoire). CEMAT is responsible for a number of important European spatial planning initiatives (see Chapter 5). Through CDAT, senior officials of member governments have provided the expertise and acted as a steering committee for spatial planning projects.

The third main CoE institution, the Congress of Local and Regional Authorities of Europe (CLRAE), was created under a charter adopted in 1994 by the Committee of Ministers with the overall aim of promoting fair and effective local democracy. Although it has quite recently reached the status of a 'third pillar' of the CoE with a strong political presence alongside the Committee of Ministers and the Parliamentary Assembly, its origins go back to the European

Conference of Local Authorities in 1957, and the Standing Conference of Local and Regional Authorities of Europe which was subsequently established.

CLRAE is a consultative body whose specific aims are to ensure the participation of local and regional authorities in the implementation of European integration, their mutual co-operation and active involvement in the work of the CoE. It meets annually in plenary session in Strasbourg and is composed of two chambers, the Chamber of Local Authorities and the Chamber of Regions. Delegates are appointed by national associations on the basis of geographical and political balance. The work of CLRAE is undertaken by a number of specialist working groups advised by local and regional officials.

CoE and spatial planning

Essentially, in the spatial planning field its actions and activities rely on the power of persuasion if they are to achieve any tangible benefit. Nevertheless, its contribution to spatial planning in Europe is considerable and deserves serious attention, including as it does the first attempts to develop a European spatial planning strategy and adoption of the *European Regional/Spatial Planning Charter* in the 1980s (CoE, 1984; see Chapter 5). It also sponsored the European Conservation Year in 1970, the European Architectural Heritage Year in 1975, the European Campaign for Urban Renaissance in 1981 and the European Year of the Environment in association with the European Commission in 1989.

Perhaps the most important role the CoE has played is in the stimulation of ideas about the form and content of European-scale spatial policy, and the reasons why it is necessary to think in these terms. It has also helped in the creation of a European planning community by providing a meeting point or forum for personal contact between central and local government officials and politicians, for many of whom participation in CoE projects was an educative process. Some of these officials would be civil servants responsible for international contact on behalf of ministers of spatial planning through the activities of CEMAT and CDAT. Many, however, would be planners from local and regional authorities which were participating in CoE activities and programmes.

Many of the national officials involved have benefited from this learning process when their countries entered into EU accession negotiations, or have since found themselves meeting former CDAT colleagues as members of the Committee of Spatial Development (see Chapter 3). Many of the local and regional officials have subsequently found the experience of the European dimension and of activities organised by CLRAE or its predecessor the Standing Conference valuable when faced with the tasks of conducting networking, European liaison and making EU funding applications on behalf of their authorities (see Chapters 7, 8 and 14).

OEEC and OECD

The Organisation for European Economic Co-operation (OEEC) and its successor, the OECD, are the product of early attempts to restructure Europe on intergovernmental lines. The OEEC was founded in 1948, initially in order to direct the distribution of funds provided by the US Marshall Aid scheme. The experience provided encouraging experience of co-operation between its 16 European member-states.

The OEEC was superseded by the OECD, created in 1961 as a worldwide rather than European association of developed countries. Membership reached 25 in 1994: all 15 EU and 4 EFTA countries plus Canada, Mexico and USA (NAFTA), Australia, Japan and New Zealand. Its aims are to promote economic development of its members and to contribute to world economic development through co-ordinated assistance from its members to developing countries. It is governed by a council on which all members are represented. Once a year this meets at ministerial level, and at frequent intervals at the level of permanent representatives. Decisions are based on consensus.

The permanent secretariat is located in Paris. Much of the substantive work of its over 200 committees and working parties is also undertaken by officials of national governments. Most of its activities fall outside the scope of this book, but from time to time an OECD project may fall within the spatial policy field.

An example from the 1980s was the 'Opportunities for Urban Economic Development' project which sought to stimulate the dissemination of the most effective local economic development concepts and offer advice to authorities in countries that were just embarking on such strategies. Research teams from cities in 17 different countries prepared case studies of local economic development projects which were presented at a conference in Venice and evaluated as a basis for disseminating advice on good practice. Several of those presenting case studies and participating in this project were from local and regional planning authorities.

Council for Mutual Economic Assistance

The Council for Mutual Economic Assistance, or Comecon, was founded in 1949 by the governments of Bulgaria, Czechoslovakia, Hungary, Poland, Romania and the USSR. The former German Democratic Republic was admitted in 1950, and Albania held membership from 1949 to 1962.

In its early years, it was not much more than a token organisation, being preeminently a political statement in the context of the cold war. An enhanced secretariat was set up in Moscow in 1960 and it did go on to achieve a number of joint development projects and bilateral agreements between its members and make a contribution to the growth and interdependence of the planned economies of the former Communist Party states (Blacksell, 1981).

Comecon no longer exists, but the legacy of the centrally planned economies of its former members, plus the entrenched mental attitude towards the concept of planning that is the consequence of having lived with this form of central planning, presents spatial planners with one of the greatest challenges to be found in 1990s' Europe.

The word planning is firmly associated in the minds of many citizens of former communist countries with central planning of the economy. Therefore, many found it difficult to grasp the sense in which the same word is used in its western sense to refer to a system of spatial planning and allocation of land uses in the context of a mixed economy by democratic local and regional authorities. In fact, many did not at first understand that any such intervention was accepted in a western 'free market' economy. Consequently, it was very difficult at first for western spatial planning advisers to get their message across.

United Nations Economic Commission for Europe

The United Nations Economic Commission for Europe (UNECE), based in Geneva, was for many years the only forum in which issues of spatial planning, urban development and human settlement were discussed, and experience shared, by government officials, professional practitioners and experts from both sides of the iron curtain. The intensity of professional contact between western Europe and the former communist countries is now very much greater, but the personnel involved on the central and eastern Europe side has largely changed from those who participated before 1989. However, the UNECE did help to ensure that perspective of the whole of Europe was not completely forgotten about by the planning community.

Among the last events to be held by the UNECE under the old order was a conference on urban and regional research and urban renewal, in Leipzig, then in the German Democratic Republic, in October 1988, and a seminar on the effectiveness of settlement planning, in London in October 1989. The basic purpose of these was the exchange of experiences of practice and discussion of the contribution of research evaluating planning and urban renewal. Proceedings tended to be rather formalised, with each national delegation presenting a rather sanitised account of its current policies and practice designed to highlight features which showed its prevailing ideology in good light, but these UNECE events did at least mean that a little more was known than might otherwise have been the case of conditions in Comecon countries.

The Nordic Council

The five Nordic countries, Denmark, Finland, Iceland, Norway and Sweden, formed the Nordic Union in 1952. This was modelled on the Benelux Union, and was intended to lead towards the formation of a customs union. Four of the governments (not Iceland) agreed a draft treaty in 1969 to set up an Organisation for Nordic Economic Co-operation (Nordec), but this was not implemented (Blacksell, 1981). Nevertheless, the Nordic Union countries have developed close cultural and political links, including removal of internal border and passport controls, and in 1991 it was thought that those not in the EU (all except Denmark) may make a group application for EU membership (Holland, 1994).

One initiative of the Nordic Council of spatial planning interest was the establishment of NORDPLAN, an institute based in Stockholm providing professional education and research in planning for all the Nordic countries. Since 1991, it has also played an important role in offering professional training programmes and advice to the authorities in Estonia, Latvia and Lithuania.

The European Communities and the EU

The title European Communities (EC) refers to the three communities created respectively by the Treaties of Paris and Rome, which have, since their foundation, merged to form the European Community, itself one pillar of the present EU along with the second and third pillars: Common Foreign and Security Policy, and Home Affairs and Justice.

The three communities are the ECSC, the EEC and the European Atomic Energy Community, known as Euratom. Another European community was proposed, the European Defence Community, for which a treaty was signed by the original six EC member-states in 1952, but this proposal did not go ahead owing to rejection by French National Assembly in 1954. A separate and looser defence association, the Western European Union (WEU), consisting of the six plus the UK, was however created in 1955. This now takes

1950	Schumann Declaration
1951	Treaty of Paris establishing the ECSC signed by governments of the six (Belgium, FR Germany, France, Italy, Luxembourg, The Netherlands)
1952	ECSC comes into operation
1955	Messina Conference (IGC)
1957	Treaties establishing the EEC and Euratom signed in Rome by the six member-states (Belgium, France, FR Germany, Italy, Luxembourg, The Netherlands)
1958	The Treaty of Rome comes into force (1 January)
1967	The Merger Treaty, creating single Commission and Council for the EC
1973	First enlargement, accession of Denmark, Ireland and UK (nine member-states)
1981	Second enlargement, accession of Greece (ten member-states)
1985	Luxembourg IGC
1986	Third enlargement, accession of Spain and Portugal (12 member-states)
1987	Single European Act comes into force (1 July)
1989	Fall of Berlin Wall, revolution in central and eastern Europe
1990	German reunification (3 October)
1991	Treaty of European Union negotiated in Maastricht
1992	Single European Market (from 31 December)
1993	European Economic Area comes into force
1993	Maastricht Treaty comes into force (from 1 November)
1995	Fourth enlargement, accession of Austria, Finland and Sweden (15 member-states)
1996	IGC to review treaties

Figure 2.1 *Chronology of the European Communities*

responsibility for the Common Foreign and Security Policy, the second pillar of the union created by the Maastricht Treaty.

Key dates and events in the formation of the EU are set out in Figure 2.1. Several of these events provide the context for spatial policy development and are referred to in later chapters. Some are discussed below.

ECSC

The Schuman Declaration, quoted above, demonstrates both the overall goal motivating those pressing for European integration and the strategy. The goal was to preserve the peace and to lock Germany into common economic structures for this purpose, and the strategy, to proceed within one vital sector rather than on a broad front, Schuman's 'limited but decisive point' was the creation of the ECSC and placed coal and steel production under the control of a supranational authority, the High Authority. The treaty creating the ECSC was signed in Paris on 18 April 1951 and quickly ratified by the parliaments of the six original members, coming into force in July 1952. The UK was invited to participate but was not prepared to accept the degree of power the supranational High Authority would have over domestic authorities responsible for coal and steel production.

The pooling of sovereignty was an essential feature of the ECSC concept, and was clearly recognised and accepted by France and Germany. As Monnet said during negotiations: 'The Schuman proposals provide a basis for the building of a new Europe through the concrete achievement of a supranational regime . . . the indispensable first principle of these proposals is the abnegation of sovereignty in a limited but decisive field.' Likewise, Chancellor Adenauer of Germany made clear in a speech to the Bundestag in June 1950 that he agreed with Schuman and Monnet, declaring 'that the importance of this project is above all political not economic' (Nugent, 1991: 35).

The central institution of the ECSC was the High Authority, with Jean Monnet as its first President. It had the power to issue decisions binding in all respects on member-states on matters concerning prohibition of subsidies, agreements between undertakings, restrictive practices and in certain cases control of prices. A Council of Ministers also existed, but the High Authority had powers that were in many respects stronger than those given to its equivalent, the Commission, under the Treaties of Rome. This has meant that, since merger in 1967 (see below), the Commission has had more autonomy when operating under the Treaty of Paris powers than when operating under the Treaties of Rome.

The ECSC was generally regarded as being an economic success in its early years when tariffs were removed and restructuring of the coal and steel industries was promoted, although it received a set-back when its proposals for restricting overproduction of coal were rejected by member-states in 1958. A spatial focus was a feature of ECSC policy more or less by definition in view of its responsibilities towards coal and steel producing regions. It also represents the nearest that the ECs come to having powers over housing since coal and steel producers at that time often owned the housing occupied by their workforce.

Improvement of living and working conditions was one of the objectives of the ECSC, and low-interest housing loans were made from ECSC funds.

EEC and Euratom

The foreign ministers of the six ECSC member-states met in Messina in Sicily in 1955 to discuss proposals for further economic integration. They resolved that

> the moment has arrived to initiate a new phase on the path of contructing Europe . . . this has to be done principally in the economic sphere, and [they] regard it as necessary to continue the creation of a united Europe through the expansion of joint institutions, the gradual fusion of national economies, the creation of a common market, and the gradual coordination of social policies.
>
> (quoted in Nugent, 1991: 41)

Although this makes no mention of spatial planning as such, the issues with which it is concerned are indicated here.

The intergovernmental negotiations which followed under the chairmanship of the Belgian Foreign Minister Paul-Henri Spaak formed the basis for the two Treaties of Rome, respectively establishing the EEC and Euratom. The UK had sent an observer to the Messina Conference. It was invited to participate in the negotiations, being a member of the WEU, and did so until November 1955 when it became clear that the intention was to move beyond creation of a loose free-trade area (Nugent, 1991).

The two treaties were eventually negotiated and signed on 25 March 1957. Following ratification, they entered into force on 1 January 1958, creating the second and third of the European Communities.

The Treaty of Rome which most concerns us is that for the EEC, creating what was in the 1960s and 1970s widely known as the Common Market. The EEC is the core of the present-day EU, and the Treaty of Rome forms the basic constitution of the EU. It has been substantially amended since by the Merger Treaty, the Single European Act, the Treaty of European Union and successive accession treaties. These, in aggregate, provide the legal basis for every policy and action of the EU.

Euratom

Euratom has been the least active and successful of the three communities. When it was formed, nuclear energy seemed poised to become the main energy source in Europe, and for security reasons it was held to be desirable to place its development under international control. Euratom was expected to lead the development of a common European energy policy.

However, assumptions held in the 1950s concerning the future importance, economic viability and political or environmental acceptability of nuclear energy have consistently been refuted since. Hence the low profile of Euratom.

EFTA

EFTA was set up largely at British initiative in 1960 as a counterattraction to the EEC. Initially, it had seven members: Austria, Denmark, Norway, Portugal,

Sweden, Switzerland and the UK. It was not intended then and for most of the subsequent period did not aspire to become anything more than a free-trade area, and rejected the concept of integrated political institutions such as those of the EC. Common policies in order to achieve integration were not therefore developed. Membership of EFTA has fluctuated, the maximum having been nine. Accession to the EU has always required resignation from EFTA, which by 1995 only had four members: Iceland, Liechtenstein, Norway and Switzerland. EFTA is now linked to the EU through the EEA agreement (see below).

Merging the European Communities

The Merger Treaty, or 'Treaty Establishing a Single Council and a Single Commission of the European Communities', was signed in 1965 and came into force on 1 July 1967. It did not merge the three communities themselves although it created a unified Commission and Council of Ministers (see Chapter 3), ending the separate existence of the ECSC High Authority.

The Single European Act

The Luxembourg IGC in 1985 negotiated the basis for the Single European Act which was signed in 1986 and came into force in July 1987. It is a wide-ranging measure which reformed the EU's institutional and decision-making structure in many important ways, made the completion of the internal market (the Single European Market or SEM) by 1992 a Treaty obligation under new Article 8A, and provided an explicit legal basis for several policy sectors of importance to spatial planning including regional policy and environment policy (see Chapters 7 and 10).

1990s restructuring

The events of 1989 and 1990, overturning the Communist Party regimes in the former USSR and Comecon countries and most vividly symbolised by the opening of the Berlin Wall on 9 November 1989, have profoundly altered the context for all European policy-making. This is true of spatial policy as much as any other sector. The period 1989–90 therefore is a key break-point in any outline of the chronology of EU policy-making.

Not everything changed as quickly as this implies, partly owing to the extent of the lead-time that many policy initiatives require, and partly because the full significance of these events could not easily or quickly be taken in. The EU was effectively enlarged through German reunification in 1990, although there was no accession treaty. Instead, an existing member-state was regarded as having adjusted its boundaries.

Also, events important for the development of EU spatial policy, including completion of the SEM programme, creation of the EEA and the Maastricht negotiations occurred at around this time, coincidentally adding to the sense that the 1990s represent a new situation. A new spatial structure for Europe emerged, requiring new thinking and new policy development. This new structure forms the new European architecture, outlined below.

The year 1993 was the year in which the SEM came into operation. The SEM represents the culmination of the intention expressed in the Treaty of Rome of 1957, signed by the original six member-states to set up the then EEC in 1960, to create a common market in goods and services. The Single European Act 1987 amended the Treaty of Rome and laid the foundations for the programme to complete the common market by the end of 1992. A key element was the extension throughout the EU of the so-called four freedoms, free movement of goods, services, labour and capital, by the removal of all remaining tariff and non-tariff barriers (see Chapter 3). The principle of the SEM was extended territorially through creation of the EEA in 1992.

EEA

Following the Single European Act and the SEM programme, and as a means of expediting enlargement negotiations with EFTA members, the EEA agreement was negotiated in 1991 and came into effect in 1993 after a legal challenge from the European Court of Justice and rejection in a referendum by Switzerland, both of which required renegotiations. The objective was to extend the SEM, and the principle of the Four Freedoms, to all EFTA members. The resulting combined single market is known as the EEA. The EEA was seen by the European Commission as a transitional arrangement, not a permanent body, pending its members' accession to the EU. However, it may not work out like this. Like EFTA, it is becoming in a sense more of a repository for countries that have decided not to take up full membership of the EU.

It lives on with Norway's 'no' vote on EU membership in November 1994. Iceland is content to remain as an EEA country without seeking EU membership, while Switzerland originally signalled its intention to seek EU membership along with Austria, Finland, Norway and Sweden. Now it finds itself not only outside the enlargement discussion but with the added complication that Switzerland remains an EFTA member but not a member of the EEA following a 'no' vote on the EEA agreement in a referendum in December 1992. In 1995, the EEA therefore consists of 18 countries, the EU of 15 plus Iceland, Liechtenstein and Norway.

Treaty of European Union

The Treaty of European Union (TEU), or Maastricht Treaty as it is widely referred to, is the foundation of the whole subject-matter of this book. It is a framework treaty which sets out general objectives and guidelines, leaving the Commission to propose concrete measures in the form of secondary legislation. The TEU sets out the fundamental objectives and duties of the EU in Article 2:

> The Community shall have the task, by establishing a common market and an economic and monetary union and by implementing the common policies or activities referred to in Articles 3 and 3a, to promote throughout the Community a harmonious and balanced development of economic activities, sustainable and non-inflationary growth respecting the environment, a high degree of convergence of economic performance, a high level of employment and of social protection, the

> raising of the standard of living and quality of life, and economic and social cohesion among Member States.
>
> (EC Council, 1992: 11–12)

Several of the tasks set out in Article 2 have spatial aspects and are the subject of later discussion, among them references to balanced development, environment and economic and social cohesion.

The Maastricht Treaty also had another fundamental objective, provision of the legal basis for economic and monetary union (EMU), or the single currency. Achievement of each of these objectives was recognised to be to some extent dependent upon the other.

These two elements were the product of two separate IGCs, one for political union and one for monetary union, convened prior to the Maastricht Summit in December 1991. IGCs of this sort are not single conferences but a series of meetings and negotiations taking place over a period of months prior to the European Summit at which agreement over the final text is sought. The 1996 IGC will follow this pattern, with the added uncertainty that it may continue well into 1997.

The term 'European Union' came into use from the date on which the treaty entered into force, 1 November 1993, because it is a union of the three 'pillars' referred to earlier, the European Community, Common Foreign and Security Policy, and Home Affairs and Justice.

EU spatial policy sectors were substantially enhanced by the Maastricht Treaty. This treaty also further developed the EU's decision-making powers and created a new political forum which is likely to take a close interest in spatial policy, the EU Committee of the Regions (see Chapter 3).

Another central concept introduced in the treaty is that of a common European citizenship. One manifestation of this should be removal of internal passport and border controls. This was in principle agreed as early as 1985 in the Schengen Agreement, so called after a small town in Luxembourg which the ministers were passing in a boat on the River Moselle as they reached agreement. The Schengen Agreement provides for complete removal of passport and customs controls for travellers crossing internal borders within the EU. Schengen came into operation in 1995, but only seven member-states, Belgium, France, Germany, Luxembourg, The Netherlands, Portugal and Spain, agreed to participate in the first phase. As Chapter 8 points out, it does have local spatial planning consequences.

Central and eastern Europe (CEE)

If this book had been written before 1989, it would have been possible to get away with a simple division into west and east Europe. Either the countries east of the iron curtain would have been ignored, or it would have been necessary to refer to the Warsaw Pact and Comecon.

This is no longer possible. First, as the acronym CEE suggests, it is necessary to distinguish between central and eastern Europe. Secondly, the EU itself has already acquired, through German reunification, some former Comecon territory and some insight into the extent of the transformation required. For other former communist countries, western Europe and the EU represents the model to which they are aspiring. It is anticipated that the next round of enlargement will include some central European countries, several of which have explicitly declared the goal of membership of the EU and are ambitious to achieve this within a decade.

On the part of the EU, a major priority since the events of 1989 has been to establish political and economic links with former Communist Party states and assist in their transformation.

As Monnet (1978: 432) observed with typical pragmatism:

> unification of Europe, like all peaceful revolutions, takes time – time to persuade people, time to change men's minds, time to adjust to the need for major transformations. But sometimes circumstances hasten the process, and new opportunities suddenly arise. Must they be missed simply because they were not expected so soon?

He could not have been more precise in his anticipation of the attitude of Chancellor Kohl and the German Federal government to the historic opportunity presented in 1989. There was no treaty of accession because there was no new member-state. Instead, an existing member-state adjusted its boundaries by adding the so-called new *Länder* (states) of Germany.

A distinction between the processes of transition and transformation must be made. Transition refers to the formal institutional changes necessary for the introduction of democracy and a market economy. Transformation requires not only this transition but also achievement of the behavioural and structural changes necessary to function in conformity with the principles of democracy and a market economy. Transition from being, as the German Democratic Republic, a separate Communist Party state to membership of the EU took under one year: transformation has some way to go still.

PHARE and TACIS

The EU has set up two large financial and technical assistance programmes to assist the CEE countries: PHARE and TACIS. PHARE is the acronym for Pologne et Hongoire Assistance pour la Réstructuration Économique. It also is French for a lighthouse, suggesting the metaphor of a beacon or guiding light for the east! More prosaically, the acronym TACIS stands for Technical Assistance for the Commonwealth of Independent States.

PHARE is directed to the countries of central Europe, and excludes all the former USSR other than the three Baltic states of Estonia, Latvia and Lithuania. Conveniently for the acronym-writers, Poland and Hungary were the first two beneficiaries, closely followed by the then Czechoslovakia. The PHARE grouping includes all those CEE countries that have, or are negotiating, formal association agreements with the EU which explicitly acknowledge the goal of eventual accession to the EU. At the core of the PHARE group is the Visegrad group, originally consisting of Poland, Czechoslovakia and Hungary, who met in Visegrad in Hungary in 1991 to agree a joint approach to the EU. There are now four Visegrad countries since the Czech Republic and Slovakia split in 1993.

The agreements, known as 'Europe agreements', are linked explicitly to actual accomplishment of economic and political reform, and are intended to form the basis of a long-term relationship with the EU leading to adoption of the four freedoms, EU environmental and regional development policies and a role in decision-making structures, but not necessarily monetary union (Kramer, 1993).

Hungary and Poland have since tabled applications for accession to the EU, and the Czech Republic and Slovakia have also signalled their intention to apply for full membership. Members of the Visegrad group were invited to attend the European Summit in Essen in December 1994 and are optimistic of achieving EU membership by the beginning of the new century. Romania has also tabled an application for EU membership, while Slovenia and the three Baltic states of Estonia, Latvia and Lithuania also hope to follow the Visegrad path to EU membership.

TACIS is, as the acronym suggests, directed at the countries of the Commonwealth of Independent States (CIS) of the former USSR. Unlike the Visegrad/PHARE group, this grouping has no formal acknowledgement of eventual EU accession written into agreements, although it is not ruled out in order to offer encouragement to face the much greater problems of transformation. The 'Partnership and Co-operation Agreements' with the EU that Belarus, Kazakhstan, Russia and Ukraine have signed are trade agreements, not free-trade deals (Kramer, 1993).

New European architecture

The 'old' architecture was the situation when Europe (or at least most countries) were either in the EU of 12, or in EFTA, or in Comecon; and when the boundary of western Europe, i.e. the iron curtain, was for practical purposes the limit of west European economic space, and therefore of jurisdiction over domestic policy-making. In this context, whatever eastern limit to the geographical concept of Europe that may be conventionally accepted (for example, the Ural Mountains) was of little practical consequence.

The new European architecture is therefore taking shape with three main building blocks:

1. The EEA – countries now in the EU or EFTA forming a western crescent of the EU from Greece via western Europe to Finland.
2. Central Europe or PHAREland – the Visegrad countries, countries benefiting from PHARE with close political, economic and infrastructure links to EU member-states, and aspiring to full EU membership early in the next century under their 'Europe agreements'. Estonia, Latvia and Lithuania are the only countries of the former USSR in this category.
3. Eastern Europe or TACISland – the countries of the former USSR in the CIS, sometimes also referred to as the NISs (newly independent states), which are benefiting from the TACIS programme but have a much longer road to follow in their transformation process. There is considerable ambiguity on the part of both the EU and the NIS countries themselves regarding any question of eventual EU membership.

Some of the boundaries between these groups may not be certain yet, and some countries may fall outside these broad blocks, but the macroscale political geography of Europe can usefully be simplified in this way. EU norms, such as environmental protection standards, funding procedures, access to markets, etc., can more appropriately be applied to the PHARE group than the TACIS group since they correspond to norms that they are explicitly aspiring to, as a matter of government policy.

These three groups are of course not homogeneous. Within the first group, based on the EU, there is not only the distinctions among EU, EFTA and the EEA but also the division within the EU between Schengenland and those countries, such as the UK, which continue to maintain passport controls. In the event of a single currency being implemented in 1999, it is likely that only a selection of member-states will participate, at least at first.

Limits to the EU and to Europe

One consequence of the 1995 enlargement is that the EU has moved closer, politically as well as geographically, to postcommunist central Europe. Austria has close ties with Hungary, Sweden and Finland with the Baltic states and in the latter case with Russia, for example. Finland is in fact in a pivotal position, having the only EU land border with Russia or TACISland.

Consideration of the third grouping, TACISland, raises the question of the limits to Europe. Before any future applications for accession can be accepted, the EU must decide where Europe ends and achieve congruence between its political and geographical definition. The Ural Mountains are traditionally held to mark the boundary of Europe but have no practical or jurisdictional significance in this context. If Russia is a European country, which architecturally, linguistically and culturally it is, then Europe has a Pacific coast and connects to the Pacific rim.

Merritt tackles this issue by posing the rhetorical question, 'Should the EC perhaps contemplate a Community that one day stretches from the Atlantic to the Bering Straits?', arguing that the former USSR can neither be left outside an enlarged EU nor can it be integrated, and quoting the first EC ambassador to the USSR as envisaging in the long term two enormous economic blocks, separate but with close connections (Merritt, 1991: 44–5).

It is possible that the border between PHAREland and TACISland, such as the River Bug on the border between Poland and Belarus, may become the effective border between western and eastern European economic areas just as the line of the Rivers Oder and Neisse, marking Poland's western border after 1945, came to symbolise Europe's postwar frontiers. So the River Bug on Poland's eastern border may come to symbolise the border between PHARE-land and TACISland, dividing Europe into eastern and western economic areas (Rosciszewski, 1993).

NAFTA and the USA

The USA is the classic model for those whose concept of European integration, as expressed in some of the postwar rhetoric, is that of creating the United

States of Europe. However, in some respects, NAFTA rather than the USA may in due course come to be seen as the comparator of the EU. NAFTA is a newcomer, having received the approval of the US Congress in 1993.

One problem posed for spatial planning by the example of the USA as the model for the EU is the lack of any USA-wide spatial policy comparable to that envisaged in the EU Commission's *Europe 2000* and *Europe 2000+* studies (Commission, 1991; 1994a). However, the spatial impact of development pressures stimulated by the NAFTA agreement, especially close to the Mexican border but also between different regions of the USA, is causing some observers to call for the development of a spatial policy (Meyer *et al.*, 1995: 196–7). The EU must be the model for such a policy.

Conclusion

The EU is unique among supranational bodies in that it is a jurisdiction. It follows the federal model rather than the confederal model. Therefore, it is qualitatively different from any other supranational association of member-states, such as the CoE, the OECD or NAFTA, all of which must rely on their members' governments to adopt and implement any policies they may have agreed to pursue.

The unique feature of the EU is the extent to which it is a supranational government and jurisdiction: European law takes precedence over national law enacted by the national parliaments of member-states, and has direct effect. This means that it is applicable to the individual citizen or legal body directly, and not simply to national governments. Although the EU does not equate with Europe as a whole, for certain purposes its policy regime extends beyond its boundaries to include EFTA countries through the EEA agreement. From its foundation as the EEC, the present EU has passed through phases of being a free-trade bloc, a customs union and a single market. The ambition of some, and fear of others, is that the next phase will be that of a federal Europe with a single currency.

3

The institutions and powers of the EU

The EU is a unique form of supranational government, being more than an intergovernmental association of independent countries, as the CoE is, but less than a federal nation-state such as the USA. As Chapter 2 has shown, the EU is a jurisdiction with its own institutions of government and the power to enforce its laws and policies. It is necessary for these institutions to have the power to act on their own authority if the objectives of European integration underlying the EU are to be achieved. Attention now turns to the powers and duties of these institutions in government and policy-making.

This chapter outlines the structure and operations of the key institutions, the forms of legal instrument and the policy-making process. Emphasis is placed on those institutions with which those engaged in spatial planning and policy-making are most frequently in contact, but the other major institutions less central to the specific theme of spatial policy-making are outlined because they are nevertheless important, and anyone operating in the spatial policy field may need to know of their role and function.

The structure of this chapter loosely follows that of the EU law and policy-making process itself. At its most basic, proposals for EU law and policy are formulated by the Commission, which is considered first. Proposals cannot become law unless they are adopted by the Council of Ministers, discussed next. These two bodies are therefore of paramount significance.

As the frequently quoted aphorism has it: 'The Commission proposes; the Council disposes.' Before any proposal can be adopted, there are several stages of consideration that they must go through. EU bodies whose participation is an essential part of this process, such as the Council of Permanent Representatives (Coreper), the European Parliament and the Committee of the Regions, are considered next, followed by the other EU institutions.

There is normally extensive discussion with non-government bodies, lobbyists, expert and representative groups, plus discussion with national governments, in any sector of EU policy. In the case of spatial policy, certain official advisory and support bodies such as the Committee for Spatial Development and the Co-operation Network of Spatial Research Institutes have been set up. This chapter shows where these and other official advisory bodies, expert groups and lobbyist fit into the decision-making process, but more detailed discussion of their significance comes later.

Finally, the EU institutions can only exercise those powers and duties assigned to them by the treaties following ratification by all member-states. Any extension of competence is dependent on revision of the treaties. Every action must be based on an explicit power or legal basis conferred by an article of the treaty. The EU is said to have competence over those policy sectors for which powers are assigned in the treaties. Although competence over several components of spatial policy is now well established (see Chapter 5), the position relating to an explicit competence for an EU spatial policy as such is less clear (see Chapter 4). This may change as a result of the IGC in 1996–7.

The Commission

The Commission is in a sense the EU's executive arm of government, or its civil service. However, its power is more than that of an administrative bureaucracy. It is an independent governmental institution in its own right, with its own political leadership bound by their oath of office to uphold the interests of the EU as a whole.

The Commission is established under Article 155 of the treaty to draft legislation to put before the Council of Ministers for adoption, and to implement the resulting legislation. It also has the duty to act as guardian of the treaties – in other words, to ensure that the goals and aspirations of the treaties are being pursued as well as to implement specific measures.

Under Article 155, the Commission also has the right to 'have its own power of decision and participate in the shaping of measures taken by the Council and the European Parliament'. This gives the Commission a political role in addition to its administrative function. Commissioners may take an active role in all three stages of the legislative process: initiation of draft proposals; debate and negotiation leading to a politically acceptable form capable of presentation to the Council of Ministers for adoption; and implementation.

The term European Commission is used to refer to two distinct bodies of people. In the strict formal sense, it refers to the body of commissioners appointed by the member-states to take political responsibility for proposing and implementing EU policy. In the second sense, it refers to the Commission staff as a whole. These are considered in turn.

Membership of the Commission

The members of the Commission are appointed by member-states. Up to now, two have been appointed from each large member-state (France, Germany, Italy, Spain, the UK) and one from each of the others. Since the 1995 enlargement, this makes a total of 21 commissioners. Commissioners are normally experienced politicians, often from the respective ruling parties with experience of holding ministerial office. In the case of member-states with two commissioners, typically one is from the governing party and one from a major opposition party. The political balance of the Commission as a whole is therefore in the hands of the member-states.

One commissioner is appointed president, usually by agreement at a European summit (see below), and two are designated vice-presidents of the Commission.

Under the Maastricht Treaty, the Commission has a five-year term of office and its membership must be approved by the European Parliament following hearings, analogous to Senate hearings for US government appointments, at which nominees are questioned by MEPs. Previously, the Commission had a four-year term, and there was a two-year Commission for the years 1993–4 in order to adjust the schedule to coincide with the five-year term of the European Parliament.

On appointment, commissioners take an oath of office in which they commit themselves to perform their duties in complete independence, in the interests of

Name	*Member-state*	*Portfolio*
Jacques SANTER	L	President
Martin BANGEMANN	D	Industrial Affairs, Telecommunications
Ritt BJERREGAARD	DK	Environment
Emma BONINO	I	Fisheries
Leon BRITTAN	UK	Vice-President, Industrial Relations
Hans van den BROEK	NL	External Relations, CEE and CIS
Edith CRESSON	F	Science, Research, Education
Franz FISCHLER	A	Agriculture, Rural Development
Padraig FLYNN	IRL	Employment, Social Affairs
Anita GRADIN	S	Immigration, Justice, Financial Control
Neil KINNOCK	UK	Transport, Trans-European Networks
Erkki LIIKANEN	SF	Budget
Manuel MARIN	E	Vice-President, External Relations
Karel van MIERT	B	Competition
Mario MONTI	I	Internal Market
Marcelino OREJA	E	Relations with EP and member-states
Christos PAPOUTSIS	GR	Energy, Small Business, Tourism
Joao de Deus PINHEIRO	P	Development Aid
Yves-Thibault de SILGUY	F	Economic and Monetary Affairs
Monika WULF-MATHIES	D	Regional Policy, Cohesion Fund

Note
Correct as at September 1995.

Figure 3.1 *Members of the European Commission*

the EU as a whole and not take instructions from any government, vowing 'neither to seek nor to take instructions from any Government or body' in carrying out their duties. The wording of the oath adds pointedly that all member-states have undertaken to respect this principle and not 'seek to influence Members of the Commission in the performance of their task', a promise not always observed in practice (Owen and Dynes, 1993: 36).

A new Commission took office in January 1995 under the presidency of Jacques Santer, the Luxembourg Commissioner, for the period 1995–9. The membership of the present Commission, with their portfolios, is listed in Figure 3.1. The commissioners whose responsibilities most closely relate to the subject of spatial policy are Monika Wulf-Mathies, Ritt Bjerregaard and Neil Kinnock.

Although nomination to membership of the Commission is in the hands of member-states, allocation of portfolios is a matter for the President to negotiate with the other commissioners. This is a task requiring tough diplomacy. In spite of the oath of office, great importance and national prestige is attached by member-states to the portfolio allocated to 'their' commissioner(s). Now that there are 21 commissioners, this presents a bigger problem than previously because there really are not enough major jobs to go round. It is likely for this reason that the question of reducing the size of the Commission will be discussed at the 1996 IGC, since the problem will become impossible to solve satisfactorily if enlargement to include the applicant countries from central Europe is to go ahead with the Commission formed on the present basis.

The members of the Commission operate as a college. They meet as a body each week, usually on Wednesdays. Proposals for EU legislation brought forward by individual commissioners must be agreed at these meetings before they can be transmitted to the Council of Ministers.

Each commissioner has a team of personal advisers, known as a cabinet, whose task is to keep their commissioner fully briefed on all aspects of EU policy and on issues affecting his or her portfolio, and to act as a think-tank initiating and developing policy initiatives and strategic thinking in their commissioner's area of responsibility. Members of a commissioner's cabinet are normally from the same member-state as the commissioner and often are political advisers or senior civil servants with whom the commissioner has previously worked.

Commission administrative structure

In its second sense, the term Commission refers to the administrative structure as a whole. In this sense, the Commission is divided into 24 directorates-general plus a number of specialist services. The total number of staff, even after enlargement, is still no more than the number of employees in the municipal administration of a typical regional city. At present, the Commission staff numbers around 15,000. These can be conveniently divided into three categories: 1) the policy-makers and administrators in grades A and B, just under 50% of the total; 2) the support staff, clerical, secretarial, reception, security, etc., in grades C and D, accounting for around 40% of the total; and 3) the translators and interpreters, just over 10% of the total.

The latter category normally of course would have no equivalent in municipal or most national administrations, but is obviously essential in a multilingual organisation with 11 official languages. Multilingual operation, and the implications for spatial planning issues, are discussed in Chapter 4. For most people working in local and regional planning, it is the staff in the first category, the A and B grade officials in the official classification, from the appropriate directorates-general with whom working relationships would need to be built up.

The 24 directorates-general (DG) are each given a number by which they are always referred to. The full list is given in Figure 3.2. Anyone in regular contact with the Commission soon becomes familiar with the numbers of the DGs responsible for their particular interests.

DG I	External Economic Relations
DG II	Economic and Financial Affairs
DG III	Industry
DG IV	Competition
DG V	Employment, Industrial Relations and Social Affairs
DG VI	Agriculture
DG VII	Transport
DG VIII	Development
DG IX	Personnel and Administration
DG X	Information, Communication, Culture and Audiovisual Media
DG XI	Environment, Nuclear Safety and Civil Protection
DG XII	Science, Research and Development
DG XIII	Telecommunications, Information Market and Exploitation of Research
DG XIV	Fisheries
DG XV	Internal Market and Financial Services
DG XVI	Regional Policy and Cohesion
DG XVII	Energy
DG XVIII	Credit and Investments
DG XIX	Budgets
DG XX	Financial Control
DG XXI	Customs Union and Indirect Taxation
DG XXII	Human Resources, Education, Training and Youth
DG XXIII	Enterprise Policy, Distributive Trades, Tourism and Co-operation
DG XXIV	Consumer Policy

Note
Correct as at September 1995.

Figure 3.2 *Directorates-general of the Commission*

The DG whose responsibilities are most central to the theme of this book is DG XVI, Regional Policy and Cohesion. Several others have responsibilities which are also of importance in the context of spatial policy, for example DG XI, Environment, and DG VII, Transport.

Most of the titles are self-explanatory, but some are in Euroenglish (see Chapter 4) and need to be interpreted carefully. For example, DG VIII, Development, is concerned specifically with aid to third-world countries with which the EU has agreements under the Lomé Convention (sometimes referred to as ACP countries, i.e. Africa, Caribbean and Pacific). The work of DG VIII therefore falls outside the scope of this book, although the title may suggest otherwise. Another example that may cause misunderstanding is Social Affairs in the title of DG V. This should not be taken to refer to social welfare, over which the EU has no powers or competence. Instead, it refers to relations between the social partners, i.e. employers and employees, and the operation of the European Social Fund, which is discussed in Chapter 7.

The titles and responsibilities of each DG do not necessarily correspond with those of any one commissioner. In fact, this is the exception rather than the rule, since the number of commissioners is dependent upon the number of member-states and the allocation of portfolios is a delicate exercise in political balance. The number of staff in each DG is commonly between 150 and 450, although those for core EU functions (e.g. DG VI and IX) are larger and new DGs may be much smaller.

It should also be noted that DGs are sometimes created to achieve specific objectives which apply for a defined period only, or to undertake functions previously the responsibility of a taskforce or service. Lists of titles and numbers published at different dates may therefore vary. For example the number DG XXII was for a time given to the DG with responsibility for achieving co-ordination of the structural funds (see Chapters 5 and 7). Now it is assigned to the DG which has taken over functions formerly carried out by the Taskforce for Human Resources.

The Commission is a totally integrated administration in the sense that officials are appointed from all member-states. It is conventional that the director-general of each DG is from a member-state other than that of the commissioner. Following an enlargement, staff from the new member-states are integrated into all levels of the Commission. Therefore, some reorganisation was necessary following the coincidence of enlargement and a new Commission in 1995.

The policy-making and administrative staff are not all permanent Commission civil servants. Several are appointed on a temporary basis for a few years from member-states' national civil services. This is commonly regarded as an important step in a career. It is also possible to make such appointments from local and regional government, and some local authority planning professionals have worked in DG XVI and DG XI on this basis. Another group of temporary staff are the *stagières*, usually young graduates who are accepted on to the Commission's work experience programme (see Appendix II).

The Commission has developed its own working culture. As would be expected, it is quite different from national administrations because of its multinational composition and its distinctive role. However, the influence of the original six member-states on its culture can still be discerned. The DGs are not conventional hierarchies in which each senior rank takes overall responsibility for the work of all the more junior officials reporting to them, in a pyramid structure. Instead it is more of a cell structure in which tasks are allocated to an official or a small group, who report direct to the director or their head of division, who in turn may report to the commissioner or a member of the commissioner's cabinet.

There are different ranks of staff appointment and salary grade, depending on levels of qualification and experience. Tasks may be allocated to officials of greater or lesser experience and qualification depending on the level deemed necessary for that particular task, and perhaps on the basis of competence in the languages likely to be used for preliminary working documents and discussions. Under the cell-structure principle it does not necessarily follow that if a relatively junior person is undertaking a specific function then the immediate superiors in status will be supervising or even be fully informed about that task.

This has implications for anyone lobbying or negotiating with the Commission, as the ploy of going over the head of the responsible official to talk to the boss may misfire.

Another feature of the Commission's way of operating is an open-door policy. Clearly there are practical limits to access, but in principle officials are always prepared to agree to meet anyone who has a legitimate interest in a subject for which they are responsible. In many cases this is of great benefit from their own point of view because they need to develop an understanding of conditions in all member-states when drafting policy and legislation and it is inevitable that any one official will not know all member-states equally well. The Commission as a whole has a duty to ensure that the views of both governments and interest groups are taken into account before presenting proposals to the Council of Ministers.

Many representative bodies and lobbying organisations naturally want to take advantage of this situation and have established offices in Brussels for this purpose. Some of those concerned with spatial planning are discussed in Chapter 8. For the reasons given above, it is essential to know exactly who is responsible for your particular interest, and it is the job of offices such as these to be able to advise the people they represent accordingly. Indeed, Brussels is sometimes said to be following the example of Washington DC as a city of lobbyists, a model that is not entirely unwelcome to the supporters of European integration.

The Council of Ministers

European law is not made by the Commission. That function is the responsibility of the Council of Ministers. This institution is in a sense therefore the supreme legislative authority of the EU. The Council does not meet solely to adopt European law: it also plays an important political role in identifying new directions for EU policy-making and in stimulating policy development.

Membership

The Council of Ministers always consists of one member from each member-state, but its actual membership varies in accordance with the subject-matter on the agenda. When the Council is sitting as the Environment Council, i.e. with proposed environmental legislation to consider, it consists of the respective national ministers for the environment. Likewise, ministers of agriculture attend when the business is agriculture policy. Sometimes the title of certain councils may seem misleading. For instance, the Social Affairs Council is responsible for matters concerning the social partners, i.e. employers and employees, and is therefore attended by employment ministers. Councils with sectoral policy agendas such as these are known as technical councils.

The General Council, which consists of foreign ministers, has a wider brief including policy co-ordination, external relations and political co-operation, preparation for the IGC and other politically sensitive matters. European councils, consisting of heads of government, are no different in law from other councils, but have a high political profile and are discussed separately below.

There is no spatial policy council as such although there is a regional policy council which would discuss *inter alia* the European Regional Development Fund. Ministers responsible for spatial planning (who may be the same people as the members of the regional policy council) have however met regularly since 1989. Since 1993, these meetings have taken during each presidency as Informal Councils of Ministers responsible for Regional Planning. Their meetings have been instrumental in pushing forward the development of the EU's spatial policy framework, discussed in detail in Chapters 6 and 7.

Council powers

Although the Council is the EU legislature, its power is circumscribed by the fact that it can normally act only on the basis of proposals from the Commission. On receipt of proposals in the form of draft legislation, it can adopt regulations, directives and decisions, the three types of legal instrument in EU law.

Before the Council can act upon a proposal from the Commission, or enact a draft legal instrument, the opinions of various consultative bodies must be obtained, the most important of which is the directly elected European Parliament. These are discussed below.

Council cannot initiate or draft proposals itself. Under the treaty, it can in principle act only upon receipt of proposals from the Commission. In practice, ways have been found to allow it a significant initiating role. Under Article 152 of the treaty, 'Council may request the Commission to undertake any studies the Council considers desirable for the attainment of common objectives, and to submit to it appropriate proposals'. Council can also adopt opinions, resolutions, agreements and recommendations. Although lacking the status of law, these may be politically difficult for the Commission to ignore and can therefore be used to exert influence on the policy-making priorities of the Commission.

Growing EU involvement in policy spheres which are not covered, or ambiguously covered by the treaties, also provides the Council with opportunities to extend its influence. This can occur when member-states need to co-operate anyway, for instance on issues such as international terrorism and drug control, but also on more conventional issues where differences in national systems make formal legislation difficult. Such non-legal arrangements do not need to be based on Commission proposals. Some aspects of the development of spatial policy initiatives have occurred in this way. The role of the presidency (see below) in setting the agenda is important in this regard.

Voting

The Council has three different voting procedures: unanimity, simple majority voting (8 out of 15), or qualified majority voting. Proposals from the Commission in the form of draft directives, regulations or decisions (the three forms of EU law, see below) are normally adopted either unanimously (i.e. each member-state in effect has a veto) or by the qualified majority voting (QMV) procedure. Simple majorities are used only for straighforward matters concerning the conduct of the meetings themselves, and not for substantive matters such as adoption of EU legal instruments.

Since the Treaty of European Union came into operation, the great majority of environment measures are submitted under treaty powers for which QMV applies. Each member-state has a number of votes, very roughly proportionate to size. Since enlargement on 1 January 1995, voting strength is as follows (total: 87 votes):

10 Votes: France, Germany, Italy, the UK
8 Votes: Spain
5 Votes: Belgium, The Netherlands, Greece, Portugal
4 Votes: Austria, Sweden
3 Votes: Denmark, Finland, Ireland
2 Votes: Luxembourg

The threshold for a QMV majority is 70%. Since 1 January 1995, 62 votes are required for a majority under QMV out of the possible 87. A blocking minority coalition now requires 26 votes and must normally therefore have at least three large countries, or two large and two small countries in order to succeed. Previously, three countries were often enough to form a blocking minority. For proposals under the Social Protocol, for which the UK negotiated an 'opt-out' at Maastricht, 52 out of 77 votes are required for a QMV majority. If a minority exceeds 23 votes, a further attempt to reach a more widely acceptable text can be requested by the minority before the measure is adopted into law. This is known as the Ioannina Compromise, so-named after the venue of negotiations during the Greek presidency in March 1994 to meet British objections to the new QMV threshold after enlargement. The Ioannina Compromise was not, however, written into the accession treaties.

The presidency

Meetings of the Council of Ministers are normally convened by the President. The President is the appropriate minister from the member-state holding the six-month rotating presidency.

Holding the presidency provides each member-state in turn with the opportunity to exert great influence on the agenda and priorities of Council business. Duties include arranging and chairing all Council meetings, launching and building support for policy initiatives, ensuring continuity of business, and external representation of the Council on other bodies and EU institutions.

The presidency is held in rotation by member-states in turn for six months at a time. For the EU of 12 member-states, the order was determined by the alphabetical order of the country names in their own language: Belgique/België, Danmark, Deutschland, Ellas, España, France, Ireland, Italia, Luxembourg, Nederland, Portugal and United Kingdom. The three new member-states, Österreich (Austria), Suomi (Finland) and Sverige (Sweden), would have been eleventh, thirteenth and fourteenth in this sequence.

The UK thus comes at the end of the cycle, the last UK presidency being July–December 1992. The Edinburgh Summit, towards the end of the UK presidency, took some decisions that were important for spatial planning, including agreement on the review of the Structural Funds (see Chapter 7) and adoption of guidelines for the operation of the subsidiarity principle.

Ministers from the member-state holding the presidency are required to chair all Council of Ministers meetings, and can exercise considerable influence over the agenda and priorities attached to different items of business. Thus, the presidency provides each member-state in turn with an opportunity to put its stamp on Council proceedings by directing its efforts to those proposals on which it most wants an agreement to be reached.

One problem with the presidency cycle has been that whenever the EU has had an even number of member-states (which has always been the case until 1995, apart from the period 1973–80), each presidency always occurs in the same six months of the year. As some important business always occurs at the same part of the year (e.g. agricultural prices in spring, budget in autumn), the objective of rotation to share the workload is not achieved.

Additionally, the cycle does not give member-states an equal opportunity to develop their own priorities because more time is lost for holidays in the second six months. The January–June period has only one major holiday period, at Easter, while July–December has two, the August summer break and the Christmas period. Effectively, therefore, the second six-month period includes not much more than four and a half working months.

In order to compensate for these factors, in the current presidency cycle which started in January 1993 each annual pairing as determined by the alphabetical sequence is reversed. Thus the order of the current cycle is

1993 Denmark, Belgium
1994 Greece, Germany
1995 France, Spain

1996 Italy, Ireland
1997 The Netherlands, Luxembourg
1998 The UK

In 1998, the pairing should be the UK then Portugal, to complete the present cycle. However, following enlargement in 1995, a revision to the order of presidencies to incorporate the three new member-states which departs from the strict alphabetical principles followed hitherto was agreed by the Council of Ministers. The sequence which will apply from 1998, assuming no further enlargement occurs in the meantime, is

1998 The UK, Austria
1999 Germany, Finland
2000 Portugal, France
2001 Sweden, Belgium
2002 Spain, Denmark
2003 Greece

The system of six-month presidencies is becoming more unwieldly with successive enlargements, as the intervals between presidencies lengthen and the growing complexity of EU business means that the burden on national administrations can be excessive. This is therefore likely to be another issue on which there will be proposals for reform at the IGC (Lodge, 1995).

The European Council

The term 'European Council' applies to the EU's summit meetings. They are attended by heads of government and foreign ministers. European Councils were not part of the original structure of the EU, the concept being formally agreed at a summit in Paris in 1974. The formal basis was incorporated into the Single European Act. Before 1974, heads of government met whenever necessary as a Council of Ministers meeting to address major problems in the EU.

The purpose of European Councils is to take key decisions on the future course of the EU, agree on matters to be referred to IGCs, negotiate issues relating to the EU's institutions and treaties, and agree matters that cannot be resolved elsewhere and have been referred to it. The agenda is often quite flexible, but very much subject to the priorities of the country holding the presidency and hosting the meeting (Nugent, 1991).

European Councils must be convened at least twice each year in a location within the member-state holding the presidency, usually towards the end of each presidency in June and December. Increasingly in the 1990s, it has become normal for these six-monthly meetings to be in a location other than the national capital. Recent examples, besides the Maastricht Council in December 1991, include Edinburgh (December 1992), Corfu (June 1994), Essen (December 1994) and Cannes (June 1995).

Extra meetings would normally take place in Brussels, but in recent years they are also often in the country of the presidency. Specific summits are subsequently very often remembered and referred to by the name of the place in which they took place, especially if significant decisions or agreements were reached at them. Maastricht is the supreme case. It has in fact held two summits in 1981 and 1991 and the level of name recognition it enjoys is far greater than that of most provincial cities of its size, around 120,000 population.

Council of Permanent Representatives

Each member-state maintains a permanent representation in Brussels, in a sense a form of embassy to represent that country's interests within the EU. The representatives themselves form the Council of Permanent Representatives, known as Coreper from the acronym of its title in French. Coreper plays a vital role in EU business.

When proposals are submitted by the Commission to the Council, they do not go directly on to the Council agenda. In addition to reference to the European Parliament, Committee of the Regions and the Economic and Social Committee (see below), they are subjected to detailed scrutiny by both the Council secretariat and Coreper. It is their task to ensure that the proposals are legally and technically sound, and meet member-states' different national circumstances as far as practically possible. Agreement is sought by negotiation with the different national governments and other bodies concerned, so that, ideally, an agreed text which is expected to gain the necessary majority of votes can be put before the Council of Ministers. If this is not possible, texts may be put to the Council in order that ministers can determine politically sensitive aspects, which officials cannot resolve.

Staff of a permanent representation is often enhanced during that country's presidency in order to handle the workload.

The European Parliament

The European Parliament (EP) dates back to 1958 when it took the form of an assembly composed of members of national parliaments who were sent as delegates. Since 1979, the EP has been directly elected. It is thus unique, being an institution of a supranational body on which citizens are directly represented, rather than being represented by their national governments. Although its formal powers did not change in 1979, it gained in authority and independence of attitude. As a directly elected body, the Parliament is one of the key features distinguishing the EU from an intergovernmental association of nation-states. Its powers were increased by the Single European Act and by the TEU, giving it a significantly stronger role in the EU legislative process (see below) and in approving nominations for membership of the Commission.

Hearings for the latter purpose took place for the first time in January 1995, a procedure that has been likened to US Senate hearings held to confirm appointments by the US President to positions in the federal government.

Elections to the EP are held at five-year intervals in mid-June. The present Parliament was elected in June 1994, the next election being in June 1999. The same electoral system should according to the treaty be adopted in all member-states, but implementation of this provision has not been agreed. Meanwhile, the UK is the only member-state not using a system of proportional representation, as a result of which the party balance of the UK membership exaggerates the swing between the votes for the main UK parties and has a disproportionate effect on the party balance of the EP as a whole.

Since the 1995 enlargement, there is a total of 626 MEPs. The numbers per member-state are set out below:

Germany	99	
France, Italy, the UK	87	each
Spain	64	
The Netherlands	31	
Belgium, Greece, Portugal	25	each
Sweden	22	
Austria	21	
Denmark, Finland	16	each
Ireland	15	
Luxembourg	6	
Total	626	

Before German reunification, Germany had the same number as the other three largest member-states (81 each). The figures given above, apart from those for the three new members, were agreed at the Edinburgh Summit to take into account the enlarged size of Germany while offering some extra seats as compensation to the others.

The seat of the EP is Strasbourg, where one-week plenary sessions are held each month. Any additional plenaries are in Brussels, where the EP committees sit during two weeks each month. The fourth week of each month is available for party group meeting and other consultations. The secretariat is located in Luxembourg. This rather inefficient arrangement is one of the most important unresolved issues concerning the location of EU institutions.

MEPs form party groupings, in which they sit in plenary sessions. The minimum number of MEPs required to form a recognised party group, for which financial and secretarial support is available, is designed to encourage grouping from as many member-states as possible. The three main European political traditions, Socialist, Christian Democrat and Liberal, are represented by MEPs from most or all member-states, the Greens from 9 of the 15, and there are six other groups.

There are a total of 20 standing committees, each specialising in a different aspect of EU policy. Three are most closely concerned with spatial planning issues: the Committees on Regional Policy; Transport and Tourism; and Environment, Public Health and Consumer Protection. Rural Development is the responsibility of the Agriculture Committee. There is no urban policy committee as yet. The committees play a key role in drafting opinions and resolutions for presentation and adoption by the EP in plenary session. Such reports are normally drafted for the committee by one of its members with a particular interest in the subject-matter, who is appointed rapporteur. The rapporteur is in a strong position to exercise considerable influence on the eventual opinion of the EP, and therefore much sought after on contentious issues.

Under the TEU, the EP gained significant extra powers over the legislative process (see below).

Economic and Social Committee

The Economic and Social Committee, or Ecosoc, provides institutional representation for the 'social partners' of employers and employees, and for a variety of economic and social interests in different sectors of industry and commerce, the professions, consumer and environmental groups. Unlike the Committee of the Regions, it has a long history, having been specified by the Treaty of Rome. It must be consulted on proposals from the Commission in those policy sectors specified in the treaties. In such cases, the Council cannot agree the adoption of any proposal unless Ecosoc has delivered its opinion or indicated that it does not intend to issue one. It may also be requested to deliver opinions by the Commission or Council and, since 1972, it has had the power to issue opinions on its own initiative. These have on occasions proved to be of considerable political significance, a case being the own-initiative opinion in February 1989 on fundamental social rights, which led in due course to the Social Charter (Noël, 1994).

Ecosoc has a total of 222 members, distributed on the following basis:

France, Germany, Italy, the UK	24	each
Spain	21	
Austria, Belgium, Greece, The Netherlands, Portugal, Sweden	12	each
Denmark, Finland, Ireland	9	each
Luxembourg	6	

Each national allocation is a multiple of three because members of Ecosoc are divided equally into three groups: 1) employers and employer organisations; 2) employees and trades unions; and 3) independent interests, liberal professions.

Committee of the Regions

The Committee of the Regions (CoR) was established under the TEU in response to demands from several member-states that regional and local authorities should be directly involved in deliberations on proposals for EU policy and legislation. In several member-states, such authorities exercise considerable powers in their own right whether or not they have an explicitly federal constitution, and in all member-states, responsibility for implementing EU directives and regulations often falls on local and regional authorities. Thus, the CoR is an attempt to address both the democratic deficit and the implementation deficit.

Strictly speaking, the CoR is a body representative of the subnational level of government throughout the EU, not of regions and localities as such. It is noticeable, however, that in the process of selection of members of CoR, considerable care was taken in all member-states to ensure a geographically balanced delegation. Other factors were representation of the major cities or regional authorities.

There are parallels with Ecosoc, with whom the CoR had to share a secretariat and premises when it started operations. Membership also totals 222, plus 222 alternates, many of whom also play a full role in CoR business. These are allocated between countries on the same basis as for Ecosoc. At the time of appointment, all must be members of 'local or regional bodies' (Article 198a). The intention always was that they should be elected local or regional politicians at the time of appointment to CoR, although it may be argued (and was by the UK government) that the word 'bodies', and its equivalents in the treaty, *collectivités* and *Körperschaften*, could be interpreted as including appointed as well as elected local and regional bodies (Gallacher, 1995a). Once

Commission 1
Regional Development, economic development and local and regional finance
Subcommission: Local and regional finance

Commission 2
Spatial planning, agriculture, hunting, fisheries, forestry, marine environment and upland areas
Subcommission: Tourism, rural areas

Commission 3
Transport and communication networks
Subcommission: Telecommunications

Commission 4
Urban policies

Commission 5
Land-use planning, environment and energy

Commission 6
Education and training

Commission 7
Citizen's Europe, research, culture, youth and consumers
Subcommission: Youth and sport

Commission 8
Economic and social cohesion, social policy and public health
Special commission: Institutional affairs

Figure 3.3 *Committee of the Regions: commissions and subcommissions*

appointed, membership continues for five years whether or not local elected office is retained. Unlike Ecosoc, members form party political groups.

The work of CoR is undertaken by a number of commissions, as their committees are rather confusingly called. These are listed in Figure 3.3. The total number, including subcommissions, is 13. This was arranged so that each of the then 12 member-states could have one member of its delegation appointed to the chair of a commission. The thirteenth, the Special Commission on Institutional Affairs, is chaired by the President of the CoR and is primarily concerned with preparations for the IGC and any revisions that may be proposed to the role, status and constitution of the CoR.

Aspects of spatial planning relate to the responsibilities of each of Commissions 1 to 5 (see Figure 3.3), and in its broadest sense the spatial structure of the EU also falls within the terms of reference of Commission 8. Although spatial planning is listed under Commission 2, the lead role in preparing the opinion on *Europe 2000+*, adopted in July 1995, was taken by Commission 5.

Although a newcomer, the CoR rapidly asserted itself in the decision-making processes of the EU in the first year of its existence and is expected to seek an extension of its powers at the IGC. In many sectors of policy-making, including all aspects of spatial policy, the political authority of the CoR is expected to ensure that its opinions are influential, and it may come to be acknowledged as a more powerful body than Ecosoc.

Other institutions and EU bodies

There are a number of other bodies of actual or potential significance in the context of spatial policy. Some, with the status of an EU institution, have functions central to the whole concept of the EU. Others are more specifically sectoral.

European Court of Justice

European law is enforced by the European Court of Justice (ECJ), which therefore has authority over national courts. Its rulings are legally binding in all respects, and may therefore be equated to regulations, although national legislation may on occasions be required in order to bring national law into line with a European Court ruling.

The ECJ should not be confused with the European Court of Human Rights (see Chapter 2). The latter is a Council of Europe institution, responsible for hearing cases brought under the European Charter of Human Rights, and is located in Strasbourg.

Financial institutions

Assuming progress towards a single currency continues, the European Central Bank (ECB), located in Frankfurt am Main, is expected to become one of the most important institutions. The first stage of economic and monetary union (EMU) was participation in the exchange rate mechanism (ERM). The second stage started in January 1994 with the establishment of the European Monetary

Institute (EMI) in Frankfurt, with the task of preparation for the introduction of the single currency and establishment of the ECB as the central bank for the EMU currency system.

From the spatial policy point of view, the European Investment Bank (EIB) plays an important role (see Chapter 5). This is an autonomous body established in 1958 under the Treaty of Rome, created to facilitate the financing of capital projects which contribute to European integration and the development of peripheral and poorer regions. The EIB has supported a wide variety of transport infrastructure, energy, telecommunications and regional development projects. A substantial allocation was made to the EIB at the Edinburgh Summit of 1992 for lending to projects connected with TENs (see Chapter 9). At the same time, the European Investment Fund was created to provide loan guarantees for the same projects.

The EIB also finances projects in many other countries that have agreements with the EU. Shareholders are the member-states, and finance is raised on the world capital markets. The EIB is the world's largest multilateral financial institution with top credit rating, the benefits of which, as a non-profit-making body, it can pass on to its clients.

The European Bank for Reconstruction and Development (EBRD) was set up in 1991 in London as one of the responses of the EU to the collapse of the Communist Party regimes of central and eastern Europe in 1989–90. Its purpose is to assist all the CEE and CIS countries in their transition to market economies by pomoting entrepreneurial activities and the creation of a decentralised and multienterprise private sector.

Court of Auditors

The Court of Auditors has the responsibility of auditing the EU's revenue and expenditure, not only to ensure its legality but also from the point of view of sound financial management. It produces special reports and issues opinions on matters of EU finance, in addition to its formal powers of audit. The Maastricht Treaty granted it the status of an EU institution and gave it additional duties of reporting to the EP and Council.

European Foundation for Living and Working Conditions

The European Foundation for Living and Working Conditions (EFLWC) is located in Dublin and was established in 1975 (Regulation EEC/75/1365, *OJ* L 139 30.5.75). It operates on the basis of four-year rolling programmes which identify priorities. Projects are then generally contracted out to independent research teams drawn from throughout the EU, and managed by the foundation with inputs and advice from the social partners (trades unions and employers), governments and the Commission. The main themes of the foundation's work concern social cohesion; access to employment, innovation and work organisation; human relations within the company, social dialogue and industrial relations; health and safety; socioeconomic aspects of the environment; equal opportunities between men and women; and co-ordination, exchange of information and dissemination of findings on all aspects of its work.

European Foundation, Dublin. (photo: the author)

Not all of these relate to the themes of this book, but work undertaken under the headings of social cohesion and socioeconomic aspects of the environment does feed into the development of policies to address urban and regional problems and is closely directed towards the achievement of the central objective of the Fifth Action Programme on the Environment (see Chapter 10), that of economic, social and environmental sustainability.

European Environment Agency

The European Environment Agency was in effect a new body in 1994 although the regulation establishing it (Regulation EEC/90/1210, *Official Journal* 1990) dates from May 1990. The regulation came into effect only after a site for the agency was designated, and agreement on the location in Copenhagen was much delayed. The main role of the agency, at least initially, is to assemble and monitor data, develop assessment methodologies, set up a network of national focal points, and provide the Commission with data and advice. Priority topics specified in the regulation include air quality and emissions; water pollution; soil conditions; land use; waste management; noise; chemical hazards; and coastal protection.

Committee of Spatial Development

The Committee of Spatial Development (CSD) is a relatively new body, coming into existence as an outcome of the Dutch presidency in 1991. It is not formally part of the Maastricht Treaty, having been agreed at a meeting of ministers responsible for spatial planning in Den Haag in November 1991. The Dutch wanted it to be named the Committee for Spatial Planning (Comité voor ruimtelijke planing) in order to make its purpose more explicit, but this was not acceptable to other member-states including the UK.

The CSD is intergovernmental. It consists of senior officials responsible for EU policy from national ministries concerned with spatial planning. Their task is to co-ordinate their activities in relation to EU spatial policy and implement decisions of the Informal Councils of Ministers of Spatial Planning (see Chapters 5 and 6). One key task is the preparation of the European Spatial Development Perspective (see Chapter 12), while another is implementation of the agreement to establish the Co-operation Network of Spatial Research Institutes (see Chapter 8).

Legislative procedure, Acts and instruments

Within the EU, the Community (EC) forms Pillar I of the union. The Community is responsible for adopting legislation which falls within the scope of Pillar I. There is a wide range of legal instruments and procedures available under the Treaty of Rome as amended by the TEU. Different procedures and forms of legislation operate when decisions are made in the areas of foreign and security policy or home affairs, the other two pillars of the EU. For spatial policy purposes, it is only necessary to understand Community procedures under Pillar I of the EU.

The Commission has sole right to initiate legislation on internal market issues, and normally does so on all other matters, although the EP acquired the right of initiative in the Maastricht Treaty, enabling it to request the Commission to prepare draft legislation on a topic that it believes requires EU action. Within the Commission, proposals are prepared within the relevant DG and drafted as an internal working document which is used for consultation with other DGs.

Once agreement on a proposed text is arrived at, it is adopted by the Commission as a whole and published as a COM document with a reference made of its year of publication and a serial number. This is the first public stage of the process although some invited experts and lobbyists may have been consulted on working documents. COM documents usually contain an explanatory memorandum and a text written in the legal form appropriate to the legal instrument proposed and its legal basis. This part of a COM document corresponds to a parliamentary bill in the British legislative system, and the memorandum explains what the problems are that the proposal is designed to address, and why the Commission feels that it is a suitable subject for EU action.

Once the legal text of a COM document has been translated into all 11 official languages, it is published in the 'C' series of the *Official Journal* (*OJ*). This marks the point when the process of seeking endorsement by the EP and adoption by the Council begins. Under the TEU, there are six decision-making procedures which may be followed, of which only three are likely to apply to the subject-matter of this book. These are the consultation procedure, the co-operation procedure and the codecision procedure (see Figure 3.4).

The consultation procedure is simplest, requiring one reading by the EP and transmission of an opinion to the Council, which can then adopt the measure if it is supported by the EP. The co-operation procedure involves two readings by the EP, before the second of which the Council agrees a 'common position'. If this is acceptable to the EP, the Council can then adopt the measure. The codecision procedure is an innovation introduced by the TEU in Article 189b, giving the EP new powers. The EP can veto a common position by an absolute majority. When this occurs, a conciliation committee may be convened. This consists of 24 members, half appointed by the EP and half by the Council, which meets to try to draw up a compromise text acceptable to both institutions.

The powers of the EP have been increased substantially by the TEU, as the new procedures demonstrate. Ecosoc and the CoR do not have the same powers but, on most subjects of legislation within the scope of this book, they are normally required to issue an opinion before the Council can adopt legislation.

Council has the power to adopt proposed legislation after all procedures are complete and all opinions issued, and is the only body with this power. Although not part of the formal procedure, Coreper would normally have been active in negotiations to end up with a proposal capable of achieving agreement in Council. The only circumstance in which Council does not have the

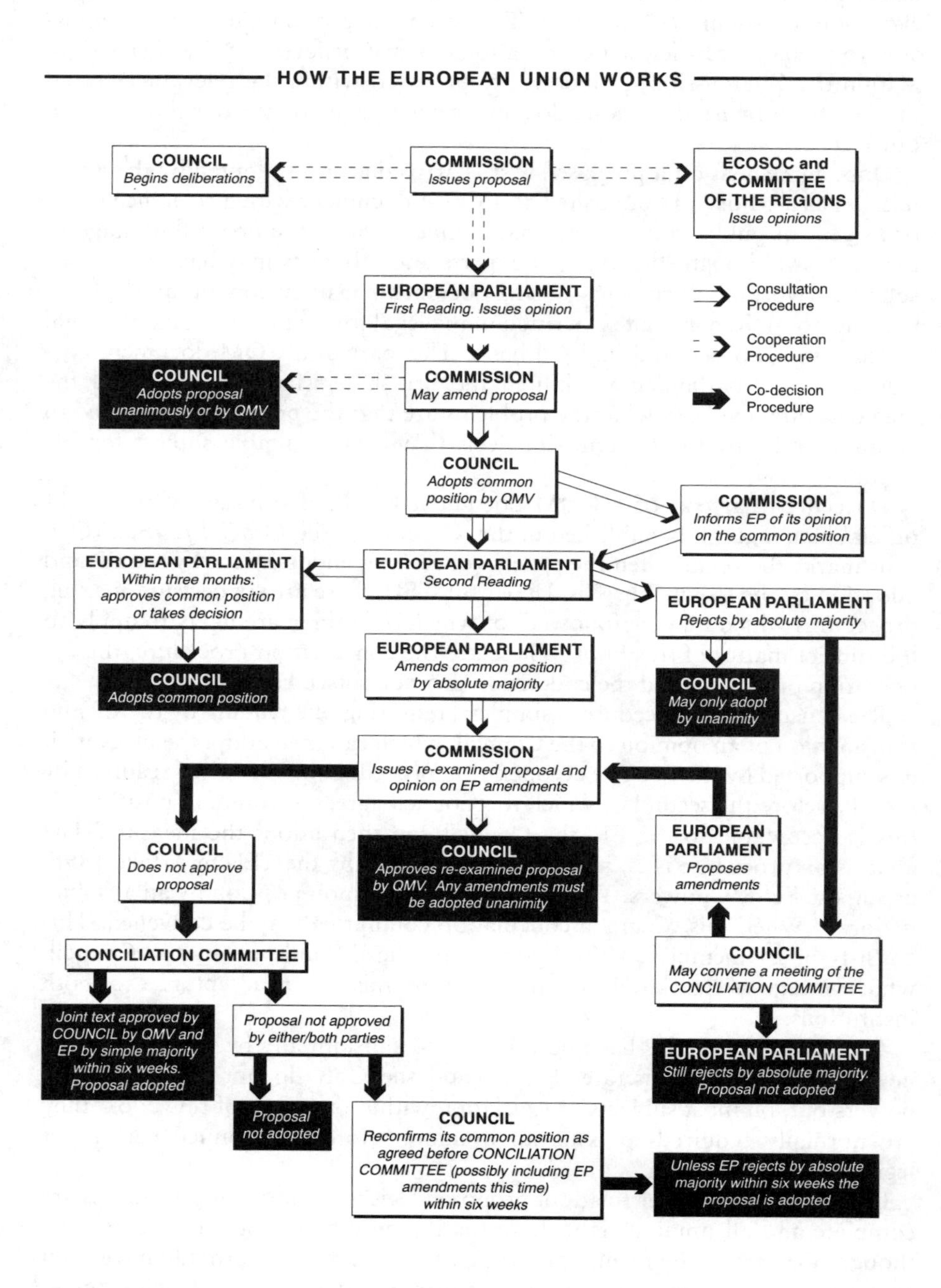

Figure 3.4 *The EU Legislation Process* (*Source*: European Information Service, LGIB)

final word is if a conciliation committee has been convened and failed to reach agreement, but the Council wishes to approve on the basis of its common position. In this circumstance only, the EP can have the last word by rejecting the proposal by absolute majority within six weeks of the Council meeting.

Once a proposal for legislation is adopted, the final text must be translated into all 11 languages and carefully checked for consistency of meaning in all languages. It is then published in the *OJ* 'L' series and notified to member-states. Unless a later date is agreed for implementation, the date on which it comes into force is the day of publication and formal notification to member-state governments, not the date of the Council meeting which may have been some weeks or months earlier. In the event of any subsequent legal action depending on discrepancies between different language versions, the ECJ is required to take the French version as definitive.

Forms of EU legal instruments

There are three types of binding legal instrument that may be adopted by the Council: regulations, directives and decisions.

A regulation is addressed to the governments of member-states and is legally binding in its entirety, i.e. the text as written enters national law directly and automatically. Regulations are normally used when it is not necessary to integrate the text with that of existing national law. The final text is published in the 'L' series of the *OJ*.

Directives are also addressed to the governments of member-states and are legally binding as to the means to be achieved, in other words the legal requirements which must be incorporated into national law are set down by European law, but the manner in which this is to be accomplished is left to the member-states. This means that the legal and policy objectives are fixed when a directive is adopted by the Council of Ministers, but the means of integrating it into national law is left to national parliaments. Normally, a period of two or three years is allowed for this process, known as national transposition, after notification and publication in the *OJ* 'L' series.

This procedure is normally adopted whenever an EU measure relates to a topic on which a body of law already exists in the member-states, as is the case with spatial planning, since these generally vary so much. It was also the procedure for the 282 measures which formed the programme for completion of the single market by 1992.

A decision is also legally binding in its entirety upon those to whom it is directed. Unlike the other two instruments, decisions are addressed to individual member-states, firms, legal bodies or private persons. Generally, their purpose is administrative, concerning implementation of existing EU law.

Compliance and enforcement

Compliance is the term used to refer to the notification required by the Commission from member-states stating that any necessary legal or administrative procedures are in place enabling a directive to be brought into operation. It is possible to distinguish between formal and practical compliance. The former is a legal process enforced by the Commission. The latter is beyond the scope of the Commission to monitor or enforce, as it does not have its own inspectorate capable of undertaking this task. The practical task of ensuring compliance and enforcement falls therefore to the authorities responsible for implementation of a directive or regulation. In the case of spatial policy instruments, this is often a local or regional authority. Under the legal doctrine that EU law has direct effect (see Chapter 2), it is possible that any legal action to enforce compliance with EU law may be pursued through the courts of that member-state.

Acquis communautaire

Whenever an enlargement takes place, the new member-states must take into their national law all the EU legislation that has been built up prior to their accession. The body of law to be acquired is referred to as the *acquis communautaire*.

In some cases, prospective members have deliberately ensured that their own legislative programme matched that of the EU in order to minimise this task. For example, this was the case with Sweden in respect of environmental legislation. The EEA agreement with EFTA was also partly designed to simplify accession in this way, especially in respect of single-market measures.

The *acquis communautaire* does not always have to be achieved by the first day of EU membership. A transition period for this purpose following accession is normally agreed in enlargement negotiations and will undoubtedly be an important feature of the next enlargement negotiations. Conversely, some parts of the *acquis* may be embodied in legislation adopted in anticipation of membership, as occurred with the EEA agreement and the 1995 enlargement countries.

Non-binding instruments

A whole range of other documents can be issued by the Commission which do not have the force of law but which may nevertheless be of major significance in setting out issues and policies, not least in the context of spatial policy.

Recommendations and opinions can be issued by the Commission, the EP, Ecosoc and the CoR to express a view on a topic or encourage adoption of good practice. Opinions are also prepared by the EP, Ecosoc and CoR as part of the legislative process either in response to a Commission proposal or on their own initiative. They will usually appear as COM documents and in the *OJ* 'C' series.

Resolutions are issued by Council and EP and are declarations of intent expressing a political view intended to establish fundamental principles for future EU action. They normally are published as COM documents.

Green papers and white papers are prepared by the Commission and published as consultation papers addressed to interested parties, whether expert groups, lobbyists, people professionally or commercially involved in an activity or anyone else. They are designed to focus on the development of policy on a subject for which the EU has not yet legislated, but may do so. Responses are usually invited within a set period. The main difference between the two is that a green paper usually focuses on one policy sector, whereas a white paper has a broader focus across several policy sectors and the EU as a whole. A good example of the former is the 1990 *Green Paper (or Book) on the Urban Environment* (Commission, 1990b; see Chapters 10 and 11), and of the latter, the Delors White Paper on *Growth, Competitiveness and Employment* (Commission, 1993b). Both green and white papers are issued as COM documents and may subsequently lead to proposals for legislation. Communications are prepared by the Commission and may follow from a green or white paper, or they may be green or white papers.

Notices are simply to give public information concerning interpretation of EU policy and law. Although not legally enforceable, they may be used in support of court cases especially in competition law. They appear in the *OJ* 'C' series.

Studies are prepared by outside bodies, such as consultants or universities and research institutes, at the request of the Commission to examine a particular subject. There is no formal publication requirement but several studies of significance in the formulation of EU spatial policy have been published by the Commission and are discussed in the appropriate chapters in Part II.

Location

The Commission is largely located in Brussels, with some offices in Luxembourg. At the time of writing, the Berlaymont building which so often is taken to symbolise the Commission is undergoing refurbishment to remove asbestos, and is expected to be reoccupied in 1998.

A summary of the location of the main institutions referred to in this chapter is given in Figure 3.5. Strictly speaking, location of the Commission in Brussels is

	Main	*Others*
Commission	Brussels	Luxembourg
Parliament	Strasbourg	Luxembourg, Brussels
EFLWC	Dublin	
EEA	Copenhagen	
ECB	Frankfurt am Main	
Council	Brussels	Country of presidency
ECJ	Luxembourg	
Coreper	Brussels	
CoR	Brussels	
Ecosoc	Brussels	
EBRD	London	
EIB	Luxembourg	

Figure 3.5 *Location of EU institutions*

provisional pending a decision on the permanent location of EU institutions. The most contentious issue is not the location of the administration but of the plenary sessions of the EP, currently in Strasbourg. In spite of this, Brussels is becoming the *de facto* seat of EU government and a lot of construction has taken place there to provide suitable buildings not only for the Commission but also for the Parliament and Council of Ministers. Another indication of permanence is that since ratification of the Maastricht Treaty it has been the policy of the Commission to purchase a freehold interest in office property in Brussels, whereas previously it rented its office space requirements. Whether leasing or buying, the EU is a very influential actor in the Brussels office market, accounting for 13% of the total office space in Brussels. International business and lobby groups attracted to Brussels because of the presence of the EU institutions account for a further 12% (de Brabander and Verhetsel, 1995: 78).

Location of EU institutions can itself be a factor in the changing spatial structure of Europe (see Chapter 6). In particular, the decision to locate the ECB in Frankfurt am Main substantially reinforces that city's role as a major international financial centre.

4

Language and meaning: the Tower of Babel problem

Why should a book about planning contain a chapter about languages? The simple answer is because the planning subject under consideration, the EU, now has 11 official languages. Austria, Belgium, Ireland and Luxembourg are the only member-states not contributing to this number. Furthermore, by reference to some of the languages other than English it is possible to argue that the EU already has a competence in spatial planning. The real question to be faced, therefore, is why does the importance of language need emphasising? Because language is a subject too easily and too often overlooked in the English-speaking world.

In order to understand the range of ideas that feed into EU spatial policy, and to be able to operate professionally on EU issues, it is vital to recognise the different forms of English encountered, to appreciate the relationships between planning thought in other member-states, the words used to express them in their languages and the dictionary translations into English. Above all, it is vital to appreciate that British-English is not everyone's base point and that its use can distort meaning in planning.

Since the TEU came into effect in 1993, this issue is of added importance. Article 130s(2) gives the EU competence over 'town and country planning' and 'land use' and therefore gives the Commission power to make proposals for legislation on these subjects, so unless everyone is clear about what the words used in each language actually mean, and how their sense corresponds one to another, there is grave danger of confusion.

In spite of the inclusion of this article in the treaty, there have continued to be calls for the EU to be given powers over spatial planning because the above article comes within the environment title of the treaty and is subject to unanimous voting. It does not therefore offer the general competence called for by the EP and CoR with the support of some national governments. It is for this reason that, when the ministers of spatial planning meet, they meet as informal councils. On the other hand, in certain language versions of the treaty, the power does appear to exist. Article 130s(2) is certainly acknowledged by the Danish Minister for the Environment, who referred to it in the introduction to the Denmark 2018 study (see Chapter 6) as one of the reasons why it was necessary to take a European view of Danish national spatial policy. German is one of the languages in which Article 130s can be construed as referring to spatial planning, but nevertheless the German government is one of those supporting calls for an EU competence in spatial planning.

Ambiguity over this one article is only one small symptom of a much wider problem. This is the whole problem of language and meaning in the vocabulary of spatial planning in the different EU languages. This chapter explores the issues of language and terminology and analyses their significance in spatial policy and planning, in order to enable the reader to understand this and to recognise the disadvantages of knowing only one language when working on EU issues or collaboratively with colleagues from other member-states, even if that one language (i.e. English) is the one that will actually be used.

It considers the prominent position of English, the different forms of English in use, Euroenglish and Eurojargon, some problems of concept and meaning of spatial planning terminology in translation, and outlines some of the ways in which the Commission and other EU institutions handle practical aspects of multilingual operation.

Language awareness is important even for those who expect to make their entire professional career within their own country and language because of the need to understand, interpret and operate policy instruments and procedures based on EU programmes and directives, for which an understanding of the underlying logic is necessary. Guidance may be written in Euroenglish; anyone may need to respond to enquiries or play host to delegations from other countries; or their firm or authority may intend to participate in a CoE or EU project or bid for a contract in another member-state.

The fundamental message is more basic: language is a cultural construct developed around the particular way of doing things that has evolved in those countries or cultures associated with that language. Therefore, to people coming from outside that country or culture, it is a filter to understanding. Because ideas and concepts are cultural products, and a nation's language evolves around the particular culture and manner of government of that nation, the words used to convey these ideas and concepts do not always lend themselves to exact translation.

This is especially evident for specialist vocabulary in an activity which has cultural parameters, which is a branch of public policy, which relates to national constitutional structures, and which exists as a result of professional practice and government action taken within countries independently of one another in response to their own problems. Spatial planning is just such an activity. Consequently, planning language and terminology is very culture specific, and can pose particular problems for the unwary.

Use of English

British organisations and authorities are very active participants in many networks and European associations in the spatial planning field. For example, almost one quarter of the member cities of Eurocities (see Chapter 8) are from the UK. Many such non-government associations have one or two official languages. Eurocities is officially bilingual in English and French, for example. Some less well resourced associations have no alternative but to adopt a single language policy for financial and practical reasons. For any association seeking to operate over the whole of the EU or the whole of Europe without the resources to pay for translation, a single language has to be chosen and the choice is almost invariably English. The Association of European Schools of Planning (AESOP) is an example. English is also the main working language of EFTA, even though it is not spoken in any EFTA member-state, for reasons of practicality and political neutrality.

Versions of English

One problem associated with the worldwide use of the English language is that various versions of it exist. For our purposes, three versions need to be identified: British-English, American-English and Euroenglish.

British-English refers to normal domestic UK usage. In fact, English usage is the basic point of reference, from which specific Scottish and Irish planning vocabulary is a variant. American-English can be quite distinct and by no means always mutually understandable. It is important in the European context because it is the language of many contributions to technical literature and also because many non-British planners learn American expressions, both general vocabulary and technical terminology. More confusingly, some continental professionals may have a mixture of American and British vocabulary without always being sure which usage is which, perhaps as a result of learning British-English at school and taking a graduate course at a US university.

Euroenglish can be defined most simply as non-British concepts conveyed using English words (Williams, 1989). It can occur whenever domestic British usage does not accurately convey the sense of the non-British original. There is a growing Euroenglish vocabulary to express ideas, concepts, policies, etc., which are neither British nor American in origin, but which are derived from other European cultural contexts, thinking patterns and the languages associated with them.

In a sense this book is written in Euroenglish. By using the term spatial planning in the title, I am using a Euroenglish term that is only slowly gaining recognition among planners in Britain who hitherto have been familiar mainly with domestic UK terminology.

Is there a particular problem with the English language?

Although English is so often the most practical language of communication in town planning, land use and real estate, it is not always the most appropriate universal language in the sense that, by being best able to express concepts from British usage and practice, its technical terminology is built up around an atypical system of planning and real estate. Furthermore, British domestic practitioners do not always realise how distinctive UK practice and terminology is.

An anecdote illustrates this very well. In response to a presentation at a conference by a British planner about a development project, a German-speaking delegate posed a question: 'Is there a plan for this project?' The answer came as a long and tortuous rigmarole about how the local planning authority had only just started work on its Unitary Development Plan, and how the structure plan notionally still applied but was out of date and not relevant.

This left the questioner somewhat confused. If the question is translated from German Euroenglish into British-English, it would have been: 'Does this project have planning permission?' This would no doubt have received a straightforward answer.

The title of the subject

Town planning is a generic term in British-English (and therefore presents no problem for British regional and rural planners when used in the title Royal Town Planning Institute) but translates as *urbanisme* (French) or *Städtebau* (German), both of which have much narrower urban planning or urban design meanings. The American generic term equivalent to the British town planning is 'city planning'. The Euroenglish word, as used here, is 'spatial planning', a term not used on a day-to-day basis in domestic UK practice although its use and recognition as a Euroenglish term is increasing. It is a valid translation of *aménagement du territoire* (French) or *Raumplanung* (German). *Raumordnung*, used in the treaty, pedantically could be taken to mean something different but in practice is recognised by German professionals as being coterminous with *Raumplanung*. The Dutch have similar words, although some Dutch professionals claim to recognise a distinction between the Dutch equivalents of the German words, i.e. *ruimtelijke planning* and *ruimtelijke ordening*.

Article 130s(2) of the Maastricht Treaty, where 'town and country planning' appears, comes within the environment title. This is arguably the wrong place if its insertion in the treaty was really intended to create a competence in spatial policy, as the equivalents, *aménagement du territoire, Raumordnung*, suggest. However, examination of the terminology used in Article 130s(2) of the

D	*Raumordnung, Bodennutzung*
DK	*fysisk planlægning, arealanvendelse*
E	*ordenación territorial, utilización del suelo*
F	*aménagement du territoire, afféctation des sols*
GR	ΠΟΛΕΟΔΟΜΙΑ ΧΩΡΟΤΑΞΙΑ ΚΑΙ ΧΡΗΣΗ ΓΗΣ
I	*assetto territoriale, destinazione dei suoli*
NL	*ruimtelijke ordening, bodembestemming*
P	*ordenamento do território, afectaçao dos solos*
S	*fysisk planering, markanvåndning*
SF	*kaavoitus, maankäyttö*

Figure 4.1 *Entry for 'town and country planning' and 'land use'* (*Source*: Treaty of European Union, Article 130s(2) p. 59, English version).

treaty for the equivalent of 'town and country planning' and 'land use' in the other 10 official EU languages provides a perfect illustration of the variety of meaning and concept of planning present in the EU. These are set out in Figure 4.1.

Sooner or later the equivalence of these terms listed in Figure 4.1 may need to be tested either for the Commission's purposes or by the ECJ. In the event of a legal dispute going to the ECJ over any conflict of meaning between the different language versions, the legal principle is that the French version takes precedence, so what spatial planners do is *aménagement du territoire*.

These terms are by no means all true translations of each other. Some terms have a more physical connotation, some connote a degree of integration between spatial and economic planning, and several convey very specific approaches to planning peculiar to their own country. This is true of both the French and the English: *aménagement du territoire* is a distinctively French concept whereas the English term is very specific to the distinctive British planning system. The German *Raumordnung* also connotes a distinctive national approach, but it is not alone. The Greek is a close translation and the Dutch is fairly similar in meaning.

Translation issues and planning terminology

The various terms used in the treaty only touch upon the problems of spatial planning terminology. Interpreting and translating legal and professional terminology in all EU languages, and ensuring that the same understanding is conveyed to speakers of these languages, is notoriously complex. The scope of planning interests and the multidisciplinary nature of the subject and its contributory disciplines add to the difficulties.

This problem was highlighted in a report on the incidence of derelict land in former coal and steel areas, discussed in Chapter 10 (RETI, 1992) where the wide variety of disciplines from which words are taken (in this example, regional development, chemistry, environment policy, property development), caused complications which could only be overcome by agreeing some common European definitions.

These issues can also be illustrated by reference to a key issue in urban planning and policy-making, referred to in British usage by the term 'inner-city policy'. In the domestic UK context the meaning is completely understood, but in a multilingual European context both 'inner-city' and 'policy' present traps for the unwary.

The German *Innenstadt* translates 'inner city' but does not distinguish between the city centre and inner areas in the way British terminology does. This is known to have caused confusion in official circles in Germany. A contrived circumlocution, *Innenstadtrandgebiet*, comes close to the geographical sense of inner city. *Stadtinnenpolitik* may be a better translation of 'inner-city policy'.

The capacity of terminology to obscure the distinction between city centre and inner city was even better illustrated by a simultaneous translation from English to Italian of a talk on UK urban policy. 'Inner city' was translated as *centro storico*, i.e. historic centre, the logic being that city centres are almost always also historic centres in Italy. A double transposition of meaning had occurred, neatly demonstrating the cultural dimension of professional terminology.

The word policy also causes problems because the German and French, *Die Politik* and *la politique*, correspond both to policy and to politics. This distinction underlies much policy discourse in English, and its loss in other languages can cause misunderstanding.

Different interpretations of what constitutes contaminated land and how much land is affected also present terminological and linguistic problems, as both the Commission's green paper on remedying environmental damage (Commission, 1993a) and a report by a lobbying organisation, RETI, point out. The RETI report concluded that there is a serious linguistic impediment to the development of contaminated-land policy at the European level. Transfer of experience and promotion of a common understanding of the nature of contaminated land and of existing national policy instruments are made difficult by the regular generation of new vocabulary 'with a large dosage of words the original meaning of which has been changed to fit a specific use' (RETI, 1992: 37), the wide variety of disciplines from which these words are taken, and the large number of words given quasi-legal meanings for specific purposes.

Several people have attempted to prepare glossaries of planning terminology (Logie, 1986; Venturi, 1990), but this is an exercise fraught with difficulties. A simple dictionary approach to finding equivalent terminology can often fail if the user is not familiar at least to some extent with the languages concerned, the ways of thinking about planning and the technical terms adopted in the two languages involved. Those who do not have such understanding and assume that a glossary can solve any problems will sooner or later be victim of a mistake or misunderstanding that could be embarrassing. An attempt to overcome this problem by offering a sentence of explanation where simple equivalents could cause confusion is found in van Breugel *et al.* (1993).

Practical operation

Not only must final versions of all official EU documents be available in English but it is also usual for early draft and unofficial versions and working documents to be in English as well. Often at this stage they also exist in French, but possibly not in any other language, depending on the composition of the team of officials working on a proposal. Many officials are able to operate, at least to some extent, in these languages, and those outside the official process such as lobbyists are much more likely to be effective if they also operate in these languages. Drafts in these languages sometimes circulate quite widely, and finalisation of the text in both these languages is often completed well in advance of those in other languages.

This appears at first sight to offer native speakers of English great advantages, and undoubtedly is part of the explanation for high British participation rates in European associations. Other reasons can also apply, as Chapter 8 discusses. Meetings are conducted on home territory, linguistically. In the early years of British membership, some British delegations attempted to speak in French and not stick to English. This tended to be a mistake as home advantage was given away, without the tactical advantage that could come from speaking Italian or German.

The domination of English is not without its disadvantages, however, especially for the monolingual. There is the general disadvantage that in any multilingual situation a monolingual participant misses much of the informal exchanges, even if there is no impediment to more official communication. Monolingual people often fail to appreciate the effort required by others to conduct business in a foreign language. This is not just a matter of courtesy, it is also a question of appreciating the different thinking patterns, differences of meaning and potential for misinterpretations and mistranslations which may be anticipated if the listener has some knowledge of the source language. This could interfere with the ability to understand properly what is being said or meant whenever the choice of English word is not exactly correct.

Another disadvantage which occurs when conducting negotiations with delegations from other member-states is that an English-speaking team does not have its own secret language. It must always be assumed that every remark to colleagues can be understood by everyone present. Delegates from countries with less widely known languages have a better chance of conducting private discussions. This can compensate for the disadvantage for them of negotiation in their second or third language. It is not unknown for representatives of Welsh authorities to use Welsh among themselves for this reason.

Many informal meeting and discussions take place without official interpreters by means of one or two common languages. It may be supposed that more formal meetings, where official interpreters are provided, avoid many of the difficulties and inequalities that can occur without interpretation. Other difficulties can arise, however. Participants need to learn through experience how to use the headsets to speak, listen or interject, whether it is better to listen to the interpretation or rely on their own knowledge of other languages, and how to speak with the pace and vocabulary that will enable the interpreters to render your contribution as accurately as possible.

The official interpreting staff have to cope with so many specialist subjects that the exact professional terminology may not always be offered in simultaneous interpretation, and participants with a partial knowledge of another language that includes their own specialist vocabulary may find it better to listen to other speakers directly. Whenever possible, the text or draft of prepared contributions to meetings and guidance on any points or terminology requiring particular accuracy should be given to the interpreters in advance so they can prepare for what is coming.

Acronyms and Eurojargon

The Commission has been most creative in the construction of acronyms, so much so that anyone who is working regularly on EU programmes, for example for a local or regional authority, must acquire a whole vocabulary of acronyms. These tend to be used alongside EU abbreviations and Euroenglish terminology, adding to the distinctiveness of EU policy discourse and professional jargon.

Some are true acronyms, in the sense that they are built up from the constituent initial letters (e.g. PHARE and TACIS), while others are built from first syllables, for example Coreper, from its title in French, Conseil des Représentantes Permanentes. Many are loosely derived from the initials, e.g. ERASMUS, from European Community Action Programme for the Mobility of University Students. The programme of Community Initiatives discussed in Chapter 7 is a rich source of examples. The chosen title of some other programmes is a sobriquet which evokes the right image. An example is SOCRATES, the successor to ERASMUS (see Appendix II), which people sometimes seek to parse as an acronym.

There is also extensive Eurojargon, of which many examples are introduced in this book. Some are words which are already familiar, but which are given a new sense which often derives from the meaning of the equivalent word in French or another language. An example is 'competence' in the sense of having the legal power and authority to act, as distinct from the sense of being capable or skilled in something. The jibe that the Commission can at the same time be both 'competent' and 'incompetent' is a wordplay on these two senses. Another example of a Euroenglish usage is of 'transparency', in the sense that the processes by which decisions are taken are visible to the mass media and the public, or in other words that the press are free to report and comment.

Other words have not been particularly widely used, if at all, until their EU meaning has been given to them. One of the most important of these is 'subsidiarity', the word that entered EU discourse after Maastricht for the principle that power should be devolved to the lowest appropriate level of government. In the context of funding programmes, 'additionality' has become a much-used word, denoting the principle that EU payments should be additional to those from national sources. This is especially an issue with the Structural Funds, discussed in Chapter 7.

Given the historic dominance of the French language as the EU institutions were created, it is not surprising that it is also possible to identify Eurofrench contributions to Eurojargon. An important example is *acquis communautaire*, referring to the body of EU legislation which must be accepted and taken into national law by any new member-state as part of the accession process (see Chapter 3).

Although most of the jargon terminology is in common use among those with professional responsibility for EU affairs, those on the inside of the working structures or comitology of the Commission and other EU institutions may lapse into a more extreme form of Eurojargon, Brenglish or Bringlish, i.e. Brussels-English. While such extremes are avoided as far as possible in this book, no apology is made for the introduction of Euroenglish terminology and EU acronyms as the text progresses because it is central to the educational purpose that these terms become a part of the reader's regular vocabulary. A number of linguistic points are therefore made in other chapters.

PART II

Policy sectors

5

A chronology of spatial policy development

This chapter sets out the context for Part II of this book, which concentrates on the core spatial policy theme. It outlines the growth of the body of policy that now exists at EU level to which the label spatial can be attached, and shows how this body of spatial policy has built up since the signing of the Treaty of Rome.

Much of this body of policy was not initiated under any explicitly spatial title, having been formulated as part of environment, regional or other policy programmes. There is, however, a growing body of policy-making, developed for the most part in the 1990s, to which the label spatial is explicitly applied, which is only briefly outlined here.

This chapter adopts a chronological framework to provide an overall picture of the evolution of EU spatial policy, and of the evolution from a purely 'economic' EEC to the present EU with a matrix of sectoral and spatial policies. Chapter 6 considers ways in which the concept of thinking spatially about Europe as a whole has been developed, Chapters 7–11 concentrate in more detail on policy in the sectors of spatial significance, and Chapter 12 takes stock of the state of spatial thinking and policy development in the mid-1990s, prior to the 1996 IGC.

Sectoral policies, of course, date back earlier than any explicit concept of spatial policy does. So the second purpose of this chapter is to make the connection between these two sources of EU spatial policy and to acknowledge the question of whether EU spatial policy is one policy or several. We return to this question in the discussion of future prospects in Part IV. A third purpose is to draw attention to the relationships among broad spatial policy, policy for cohesion and the EU budget.

The chronological structure adopted in this chapter sets out the stages by which the components of EU spatial policy have been assembled. This will also provide a framework on which the main sectoral policies of significance for spatial planning can be located. For convenience, the treatment of the material is subdivided into five time periods which may readily be related to the overall chronology of the EU outlined in Chapter 2. They also approximate to the successive enlargements, not entirely coincidentally. The changing spatial structure of the EU is a factor underlying spatial policy development at the EU scale to an extent not normally applicable within national or regional jurisdictions (see Chapter 6), and the successive enlargements represent massive readjustments to the spatial structure of the EU. Several components of spatial policy owe their origin to this factor.

In order to compensate for the spatial consequences of accession, such as exposure of economically weak regions after protection through national tariffs is removed, acceding member-states often seek to negotiate some new policy instrument or extend an existing one. Likewise, existing member-states may take the political opportunity presented by enlargement negotiations to strengthen spatial policies which may help their problem regions to face new competition from new member-states. Therefore, several components in the package of spatial policies that has grown over the years are directly or indirectly the consequence of enlargement negotiations. Integral to the politics surrounding enlargement negotiations are budgetary considerations. This chapter also aims to demonstrate the link between spatial policy or its components and the EU budget.

Treaty of Rome

The original Treaty of Rome of 1957 included the policy sectors which, it was anticipated at the time, would play an important role in the creation of a common market. Chief among these in terms of spatial significance were agriculture and transport. In both cases, this was recognised in the treaty: Article 39, concerning the CAP, referred to the need to take into account the 'natural disparities between the various agricultural regions'; while Article 75 on transport allowed for member-states to make special provision if common measures 'seriously affect the standard of living and the level of employment in certain regions'. In addition, Article 92, which was concerned with reducing distortion of competition owing to state aid to industry, nevertheless provided for exemptions for 'aids intended to promote the economic development of regions where the standard of living is abnormally low or where there exists serious unemployment' (Nevin, 1990: 289).

The European Social Fund also dates from the original treaty, Article 123, but for many years its scope was limited to retraining and resettlement allowances, and income support for short periods of unemployment (Lodge, 1993: 153), although a disproportionate share of its expenditure was in the poorer regions during the 1960s. The spatial dimension of its activity was only made explicit in 1972 and it was only in the 1980s that it began to play a role in urban and regional development strategies.

The Social Fund was nevertheless included in the treaty partly for spatial reasons, as was the EIB. Article 130, which established the EIB, noted the considerable disparities of wealth between the different regions. Italy was concerned about the consequences for its weaker regions and social groups, and therefore obtained their inclusion (Lodge, 1993: 63).

It was also recognised that the need may arise for measures not explicitly anticipated by the treaty. Articles 100 and 235 were included for this reason. Article 235 provides that

> if action by the Community should prove necessary to attain, in the course of the operation of the common market, one of the objectives of the Community and this Treaty has not provided the necessary powers, the Council shall, acting unanimously on a proposal from the Commission and after consulting the Assembly [i.e. what is now the EP], take the appropriate measures.

This article was commonly used, for example, for environment policy measures such as the Directive on Environmental Assessment in 1985 (see Chapter 10) which predated the inclusion of an environment title in the treaty.

Regional policy

During this period there was no regional policy as such although some elements of regional funding operated within programmes that had analogous purposes. The ECSC and the EAGGF gave some financial assistance to depressed areas simply because that is where the kinds of industry they supported was located. In the case of the ECSC, this dated back to 1952. Assistance from the ECSC to areas traditionally dependent on the coal and steel industry included support for industrial undertakings, subsidies for the resettlement and retraining of redundant coal and steel workers, and conversion loans to new companies in any sector proposing to locate within coal and steel areas.

Transport policy

Transport was recognised by the authors of the Treaty of Rome to be a key sector if the ambitions of integration were to be achieved, and the basis for a common transport policy was laid down. Its fundamental objectives, as set out in Article 74, were in effect the same as for the EEC as a whole: harmonisation, balanced development of economic activities and sustainable growth. However, in complete contrast to the CAP, there was for many years almost nothing to show for it.

The lack of a coherent transport policy had become one of the greatest failures of the EU, although spatial aspects of transport policy were never completely forgotten. In a 1966 directive, for example, member-states were required to inform the Commission of any decisions to develop transport infrastructure 'of European significance'.

EIB

In a sense the most explicit recognition of the need for spatial policy in the Treaty of Rome was through the EIB. The treaty established the EIB as an independent investment bank, offering long-term loans from its own borrowings on the open international financial markets. It is a non-profit bank with a top credit rating, enabling it to borrow and relend at rates of interest lower than those which would be available to local and regional authorities in depressed regions if they were borrowing on their own account. The EIB is not an aid agency, but it is an instrument of regional development. To this end, duties and guidance in its lending policy are laid down in the treaty. Its purpose is to 'contribute to the smooth development of the common market', and it is required to give priority to 'projects for developing less-developed regions'

(Nevin, 1990: 290). The EIB is not confined to lending within the EU. One of the projects it has supported is the first bridge over the Bosphorus in Istanbul.

1970s and the first enlargement

Two major components of the spatial package of policies can be directly linked to the first enlargement in 1973 to admit Denmark, Ireland and the UK: regional policy and environment policy. Negotiations had of course included Norway, which clearly also had concerns regarding peripherality and environmental quality.

Environment

Environment policy as an acknowledged policy sector dates back to the Paris Summit of October 1972, when the decision to adopt an environment policy was taken. In fact, the earliest environment directives date from before 1973 (Haigh, 1989). For a legal basis the Council and Commission relied instead on the general provisions of Article 100 (concerning harmonisation of national laws and procedures directly affecting the functioning of the common market) and Article 235, a catch-all providing for action to be adopted by the Council acting unanimously on issues for which no provision was made elsewhere in the treaty.

The accession treaty for the three new member-states had been signed earlier in the year, so the Paris Summit was in effect a summit of nine member-states although the actual date of enlargement was 1 January 1973. In a common statement, the heads of government declared that 'economic development is not an end in itself' and that attention should be given to environmental protection in order to improve the quality of life (Wurzel, 1993: 179).

Two basic motivations led the Paris Summit to decide to launch the environment programme. First was the recognition, made explicit in the declaration of the heads of government, that concern for the quality of life and protection of the natural environment was widely felt and that the community could not be seen not to address this concern. Secondly, but less explicitly addressed in the declaration, was the recognition that if a situation were permitted to develop in which member-states took widely differing actions to deal with environmental problems and to combat pollution, there would inevitably be distortion of competition and consequently interference with the proper functioning of the common market (Fairclough, 1983).

Environment policy is presented in the form of successive action programmes on the environment (see Chapter 10). The first was agreed in 1973 for the period 1973–6. During the 1970s, the scope was limited, partly owing to the political impact of the 1973 oil crisis. It remained the responsibility of the Environment and Consumer Protection Service until it was given the status of a Directorate-General in 1981 when DG XI, then entitled Environment, Nuclear Safety and Consumer Protection, was set up. However, the second action programme for 1977–81 signalled the intention to focus on land-use planning, which it identified as the policy mechanism by which environmental objectives could be achieved (Macrory, 1983; Haigh, 1989). The Environmental Assessment Directive of 1985 (see below and Chapter 10) was a product of this focus in the second action programme.

Less favoured areas

A number of measures designed to support the economy of socially less favoured areas dates back to 1970. The Less Favoured Area Directives were in the first place based on Council Regulation 729/70 (*OJ* L94 28.4.70; Haigh, 1989: 316) concerning financing of the CAP and are thus formally part of agriculture policy, although they are also regarded as part of environment policy (Haigh, 1989) and are certainly an element of spatial policy. A number of directives have been issued, the basic framework being the directive on mountain and hill farming and farming in certain less favoured areas (Directive EEC/75/268; *OJ* L128 19.5.75).

The first action programme on the environment noted the connection between agriculture and environmental policy, stating the Commission's aim that protection of the natural environment should come within the framework of agriculture policy. The directives themselves allow member-states to introduce a system of aid for specified less favoured areas which would support the continuation of farming so that rural depopulation is averted and the countryside conserved. Three types of area could be designated:

1. Mountain areas handicapped by short growing seasons and/or steep slopes.
2. Less favoured areas or regions in danger of depopulation owing to a weak and declining agricultural base.
3. Other less favoured areas suffering from specific handicaps where continued farming is desirable in order to maintain the environment, protect the coastline and/or preserve tourist potential. Since the 1985 revision of the regulation (Regulation 797/85), this can include handicaps imposed as policy, for example through designation as national parks.

Regional policy

Regional policy is most directly associated with the accession of the UK in 1973. As indicated above, in a sense regional policy can be traced back to the original foundation of the ECSC in 1952. Its focus on the problems of the heavily industrialised coalfield regions of Lorraine in France, southern Belgium and the Ruhr in Germany was in effect an early adoption of a transnational response to spatial problems of a type instantly recognisable to anyone with responsibility for British regional policy.

Approval in principle for a regional policy as a part of the accession agreement was reached at the Paris Summit in 1972, and the ERDF was set up by the Council of Ministers in December 1974 with the objectives of 'correcting the principal imbalances in the Community resulting from agricultural preponderance, industrial change and structural unemployment' (Roberts *et al*., 1993: 41). One specific budgetary reason for creating the ERDF was to counter the acknowledged imbalance arising for the UK from the arrangements for the CAP which were expected to give Britain limited benefits (Shackleton, 1993: 101). For the then UK Prime Minister Edward Heath, the creation of a regional fund was a high priority in order to ensure tangible benefits from EU membership and to compensate for financial loss attributable to the CAP. As George (1991: 193) notes, 'all that was being sought was an institutionalised subsidy from the EC for British expenditure in the regions'.

ERDF

The ERDF came into operation in 1975. The commissioner responsible was one of the UK's first two commissioners, George Thomson. In its early formulation, a system of national quotas for allocation of ERDF funds applied, and the definition of eligible regions was in the hands of member-states. In the 1970s therefore, EU regional policy was very much the aggregate of national regional policies, with very little claim to be a 'European' policy. Apart from the operations of the ERDF itself, an integrated common regional policy for the EU as a whole was not on the agenda.

While these budgetary considerations played a prominent role in the creation of the ERDF, the link between monetary union and spatial policy was recognised even at that time. Commissioner George Thomson noted in his *Report on the Regional Problems of the Enlarged Community* (Commission, 1973) that monetary union within Europe was inconceivable without a strong community regional policy. This report proposed four criteria for selection of regions which should be beneficiaries of the ERDF:

1. Per capita income below the community average.
2. Over 20% of the population dependent on agriculture or a declining industry.
3. Unemployment persistently 20% above the national average.
4. High rate of emigration.

Although there was a strong British influence in the thinking behind the ERDF, and its role was to compensate for being a net contributor to the CAP budget, it should be noted that these criteria include many agricultural and underdeveloped regions, as well as the long-developed former industrial regions expected to benefit in the UK.

In its initial phase of operation during 1975–7, funding allocations through ERDF were made on a national basis by a system of national quotas, rather than through any EU assessment of regional needs. This was a fundamental departure from the principles laid down in the Thomson report and demonstrated beyond doubt that the ERDF was at that time in effect a budget redistribution mechanism not a European spatial policy (Nevin, 1990). Each of the then nine member-states was given a quota or percentage of the total ERDF budget available. The first set of quota allocations, for 1975–7, were as follows (%):

Italy	40.00
The UK	27.76

France	14.87
Ireland	6.46
Germany	6.34
The Netherlands	1.69
Belgium	1.49
Denmark	1.29
Luxembourg	0.10

These were not, strictly speaking, entitlements, but member-states did not need to do much more than submit a sufficient number of actual projects in eligible areas to correspond to the sum of money represented by their national quota. Projects and areas of eligibility were selected by national governments. EU regional policy was therefore no more than the sum of national regional policies at that time.

Additionality is a key word here. The Commission sought assurance from member-states that ERDF funds fulfilled the additionality requirement, in other words that they were additional to funding that would in any event have been forthcoming from national and regional sources, and were not simply substituting for them. The suspicion, to put it no stronger, that the additionality requirement was often ignored reinforced the view that ERDF was at that time primarily a budget-transfer system.

Another issue was that the vast majority of projects supported by the ERDF were of their nature very small scale. However valuable local access roads, new bridges and industrial estates may be, the Commission was concerned that they should be seen to fit within some overall strategy designed to improve regional economic performance and thereby reduce disparities between the different regions of the EU. The practical problem of assessing large numbers of individual projects for approval was also a concern (Figure 5.1, Figure 5.2).

The Commission therefore has sought over subsequent years to achieve programme funding, integration and co-ordination of funding programmes,

Region	*Industry*	*Infrastructural*	*Total (incl. other items)*
North	83.80	300.56	391.59
York & Humbs	21.56	195.46	217.44
East Mids	6.83	31.49	38.87
South West	8.35	108.03	116.63
West Mids	3.75	137.50	141.35
North West	51.81	242.06	355.80
Wales	77.79	325.05	413.52
Scotland	151.33	441.62	645.87
N Ireland	91.72	185.27	278.65

Figure 5.1 *ERDF expenditure in the UK 1975–87 (in UK £m)* (*Source*: Commission Press Release ISEC/1/88, 1988).

North	Kielder reservoir, Northumberland; electrification of east coast railway, restoration of Theatre Royal, Newcastle.
Yorks & Humbs	Transport interchange, Sheffield; tourist infrastructure, Bradford.
East Mids	Grant to aerospace and electronics industry, Castleton.
South West	Main road improvements; grants to electrical goods manufacturer.
West Mids	Relief road, Coventry; access roads for industrial sites, Birmingham and Black Country.
North West	Inner Ring Road, Wigan; grant to aluminium industry, airport development, Manchester.
Wales	Plant and machinery for 10 industrial sites; marina, Milford Haven; 18 road improvements for industry and tourism.
Scotland	Electricity cable link, Western Isles; distributor road, East Kilbride.
N Ireland	New Roll-on/Roll-off terminal, Belfast; grants to tourist and textile industries.

Figure 5.2 *Examples of ERDF-funded projects in the UK 1985–87* (*Source*: Commission press releases, 1985–88).

Europeanisation of the policy and the criteria for implementation, and a spatial policy framework within which the ERDF and other programmes should sit.

The non-quota section

A non-quota section was introduced in 1978 in an attempt to Europeanise the ERDF. The Commission was dissatisfied with the way in which the ERDF had been operating, and proposed to classify the areas benefiting from the ERDF into four categories, depending on the type of policy action needed. These were areas requiring long-term aid to overcome lack of basic infrastructure, areas requiring assistance for industries in decline, those areas adversely affected by EU policies themselves, and border areas where concerted action on both sides of an internal EU frontier by two or more member-states was required. The first two categories remained within the quota system, but the latter two were seen as EU rather than national issues for which ERDF expenditure should be allocated by the Commission according to EU rather than national criteria.

The Commission proposed that 20% of the ERDF should be made available for this purpose, but this non-quota element was reduced by the Council of Ministers to 5%. Furthermore, the Council specified that approval of projects financed from this section would require a unanimous vote. Not surprisingly, the non-quota section did not make much difference to the operations of the ERDF although some of the Commission's concerns have been reflected in later initiatives.

The Council did agree in 1978 that the Commission should give closer attention to regional measures and the regional dimension of other EU policies at both EU and national levels. Systematic monitoring procedures were put in place, criteria for identifying regional imbalance in EU terms to justify ERDF support were to be simplified and clarified, and the Commission was instructed to submit periodic reports on the state of the regions. Thus, although the non-quota section did not achieve Europeanisation of the ERDF, a preliminary step was taken in the longer-term process of establishing an EU spatial policy.

Fuller accounts of the early development of regional policy and the ERDF are given in Nevin (1990) and George (1991).

Early 1980s and Greek accession

This was a period in which steps of longer-term significance were taken in the development of EU spatial policy, in the form of conceptual thinking as well as by the introduction of new policy instruments and the adaptation of existing ones. Confrontation with the EU's new spatial structure and greatly increased disparities may have provided the stimulus for the development of the idea of a spatial policy framework for Europe as a whole, and for the first real evidence of explicit political support for such a policy.

The introduction of Integrated Mediterranean Programmes (IMPs) in 1985 (see below) was a direct response to the greater disparities attributable to Greek accession in 1981, and were agreed in order to counteract the tendency of several established policies to benefit the wealthier northern regions of the EU.

Structural Funds

The Structural Funds is the term used to refer to the three funds whose operations are now co-ordinated, the ERDF, the ESF and the guidance section of the EAGGF. Although the term Structural Funds was not in general use until the co-ordination process took place later in the 1980s, it is helpful to consider the ERDF and the ESF together at this stage.

The later 1970s and 1980s was a period during which many local and regional authorities in several countries were developing policies for urban economic development. Examples of such projects were the subject of a world-wide project by the OECD (see Chapter 2) in 1985 to review the range of initiatives and promote best practice. Within the EU, many local and regional authorities sought to utilize the opportunities offered by the ERDF to obtain financial support for their projects. Many local industrial estates, access roads, factory conversions and new factory-building projects which were undertaken as part of local authority economic development programmes at this time were part financed in this way by the ERDF.

Associated with this phase of local economic policy development was increased recognition of the contribution that the ESF could play in urban economic development. A convergence occurred, the product of both revisions to the ESF guidelines under the direction of the then Commissioner, Ivor Richard, and growing recognition of the potential contribution of the ESF in this context. ESF funding often supported training, recruitment and wage subsidy schemes which themselves formed part of local economic development packages on offer to SMEs.

As far as the ERDF itself was concerned, it continued on the same general principles as had been established before Greek accession, adjusted to give Greece a quota of 13%. Other quotas were correspondingly reduced but as the overall size of the ERDF grew, no member-state had an actual reduction in the sums of money it could claim.

A more significant revision was agreed with the adoption of new regulations by the Council of Ministers in 1984 (Regulation EEC 1787/84; see Nevin, 1990; RTPI, 1994). This was the introduction of the principle of programme funding. At the same time, the quota system was revised to give each member-state a quota range, specifying a minimum and a maximum. This system operated from 1985.

The quota ranges were cleverly devised so that each member-state, with the exception of Denmark, could theoretically obtain a bigger share of ERDF funds than had been the case before. Denmark was an exception not only because it was the richest member but also because its most backward and peripheral 'region', Greenland, had left the EU following independence in 1984. In effect, the quota ranges created a system of competitive bidding for 11% of the total ERDF available, subject to appraisal by Commission rather than national criteria.

Programme funding

Programme funding was another response to the perceived needs both to Europeanise the ERDF and to ensure that its funds were used to pursue a properly thought-out strategy for regional economic development rather than support an *ad hoc* selection of individual projects. Three forms of programme were proposed:

1. National Programmes of Community Interest (NCPIs).
2. Integrated Development Operations (IDOs).
3. Community Programmes (CPs).

NCPIs offered an incentive to member-states to present their own projects for ERDF funding within a coherent programme supported by a rationale that met EU criteria. If accepted by the Commission, ERDF funding for projects within the NPCI was approved on a multiannual basis.

An early example of an NPCI was in Shildon in the north of England. The overall objective was to co-ordinate the activities of all the organisations operating in partnership to compensate the area for the loss of its traditional employment base in railway-wagon manufacture by increasing job opportunities, encouraging private investment and improving the quality of the local environment (RTPI, 1994).

The Former Shildon Wagon Works, core of the NPCI. (photo: Sedgefield District Council, UK)

IDOs differed from NPCIs in that their key feature was that the strategy was designed to make use of more than one of the EU funds. In the more urban and industrial areas, this normally meant proposing measures which made combined use of the ESF and ERDF, while in rural areas agricultural guidance funds from the EAGGF were commonly involved. Belfast and Naples were chosen as the locations of pilot IDOs, and Strathclyde Region later established one of the larger examples in conjunction with its district councils. Strathclyde was at this time (1980–1) also one of the first UK authorities to open a Brussels office (see Chapter 8).

In the case of both NPCIs and IDOs, it was a prerequisite for funding that a regional development programme had to be prepared. In some cases, these could be integrated with regional spatial plans, but in countries where the spatial planning system required no such plan, such as the UK, they had to be specially prepared if EU funding was to be obtained.

CPs, or as they are now termed, Community Initiatives, have become one of the main elements of the funding programmes (see Chapter 7). The objective is to provide the means for a concerted attack on specific problem areas or sectors of the economy through a multiannual funding programme. CPs set up under this regulation have acquired a reputation for the ingenuity of their acronyms: early examples include the RESIDER programme for areas traditionally dependent on the steel industry, and RENAVAL similarly for shipbuilding areas.

Environment policy

Environment policy took a higher profile with the establishment of DG XI in 1981 – a notable early achievement being the adoption of the Seveso Directive in 1982. This was so called because it was proposed in response to pressure from the EP following the disaster at Seveso in Italy when an explosion allowed a toxic dust, dioxin, to escape and pollute the surrounding countryside. This directive, on major accident hazards of certain industrial activities (Directive EEC/82/501), concerns on-site and off-site plans to deal with any emergency arising from accidents occurring with the processing of specified dangerous substances (Haigh, 1989).

This period also saw the adoption of the Environmental Assessment Directive in 1985 (Directive EEC/85/337; see Chapter 10), still the most specific intervention of European law into the local land-use planning process. This directive followed the logic of the second action programme's identification of land-use planning, and the principle that prevention is better than cure. It took the process of authorisation to develop land as the means by which a procedure to ensure that development with the potential to cause pollution was assessed, and all necessary measures to minimise risk of pollution taken, before authorisation to develop (planning permission in UK usage) was granted.

This directive also had a single-market objective, to put in place the means whereby distortions of the market caused by firms taking advantage of pollution haven in member-states with weak environmental regulation could be overcome (Williams, 1986; 1988).

At that time the danger of the emergence of pollution havens creating such distortions was not fully appreciated, since the idea that one should think in terms of the EU as one jurisdiction, one market and one economic space was still not grasped by all concerned. This is well illustrated by the position taken by several organisations in Britain responding to the proposed Directive on Environmental Assessment of Projects on the grounds that the UK already had legislation allowing its objectives to be met, i.e. the town and country planning Acts, and therefore had no need of European legislation (Williams, 1983; 1986). This argument completely missed the point of the idea of a single market. Overcoming this type of misunderstanding concerning the nature of the EU was one of the reasons why the SEM programme leading up to the 1992 target date was later adopted.

Environmentally sensitive areas

Another spatial measure of an environmental nature came with the introduction of the concept of environmentally sensitive areas under Regulation 797/85 in 1985. This regulation was strictly part of agriculture policy, bringing together a number of related agriculture support measures and amendments to earlier legislation including the Less Favoured Areas Directive (see above). Under Article 19 of the regulation, member-states may introduce their own aid schemes for environmentally sensitive areas, defined as being of recognized importance from an ecological and landscape point of view, where conservation of the natural habitat should take precedence over intensification of agriculture. The type of areas designated include some areas well known as tourist areas, which subjects the landscape to greater pressure.

Spatial co-ordination

Although a body of spatial policy was beginning to be built up, it was not until the 1980s that the first essays in spatial co-ordination were prepared. Two bodies took the key initiatives, the EP and the CoE.

The Gendebien Report

The growing size and importance of the body of EU policy of a spatial nature was recognised by the EP in adopting a *Report . . . on a European Regional Planning Scheme* in 1983 (EP, 1983). This report was prepared by the EP's Committee on Regional Policy and Regional Planning. The rapporteur was Belgian MEP Paul-Henri Gendebien. Hence it came to be known as the Gendebien Report.

The resolution adopted by the EP on the basis of the Gendebien Report called for a European regional planning policy to be adopted and a planning scheme devised. Three main policy objectives were proposed. To

1. co-ordinate existing EU measures and instruments in order to ensure their 'functional and financial rationalisation in time and space' and that 'from the spatial point of view, no decision will stand in contradiction to any other';

2. 'promote balanced and integrated regional development'; and
3. protect Europe's diversity of cultural heritage and natural environment (EP, 1983: 6–7).

The resolution went on to call for the creation of an operational unit under the authority of a commissioner 'responsible for regional planning and the spatial coordination of the various Community instruments and measures' (EP, 1983: 8).

The report itself argued that European spatial planning already existed as a result of the aggregation of existing EU policies, and that in 'carrying out their responsibilities the [EU] institutions have launched *a series of policies whose respective spatial implications, when taken together, form the constituent parts of a de facto regional policy*' (EP, 1983: 11, emphasis in original). It elaborated these, reviewing the agriculture, regional, energy, environment and transport policies. It went on to set out in some detail the objectives of a European regional planning scheme, its contents and how its preparation should be undertaken. A key point was to argue that a legal basis for such a policy already existed, albeit implicitly, in the Treaty of Rome. It based this on the generalised call for balanced and harmonious development in Articles 2 and 3, and on the catch-all Article 235 (see above).

Transport

Transport was clearly identified in the Gendebien Report as a sector in which lack of action, rather than spatially unco-ordinated action, was the problem. A report from the Transport Committee of the European Parliament in 1980, calling for a comprehensive transport infrastructure plan for the EU, went so far as to note that 'following the creation of the ECSC the next step should really have been to set up a European transport community' (EP, 1980: 8).

Concern in the EP about the lack of action towards a common transport policy reached the point where the EP, with Commission support, initiated legal proceedings in the ECJ in 1984 against the Council of Ministers for its failure to fulfil its duties laid out in the Treaty of Rome calling for a common transport policy.

The ECJ ruled, in May 1985, that there was a duty to produce legislation which the Council had not fulfilled, although it was not enforceable because no fixed timescale was laid down in the treaty (Nugent, 1991). Nevertheless, the ruling was that the Council had been negligent, rejecting as completely unacceptable its defence that it had tried to develop a policy but that it had proved impossible to agree on a common policy. This was an embarrassment to the members of the Council of Ministers, and member-states did subsequently start to make progress on the deregulation of transport services and harmonisation of transport policy.

The introduction of QMV in the Single European Act (see below) improved the prospects of reaching agreement, and in 1987, a Committee of Transport Infrastructure was set up. The European Round Table of Industrialists argued at this time that a more integrated transport infrastructure was necessary if the SEM was to become a reality, emphasising the distinction between free movement as a legal right and free movement as a practical reality for goods and people. Proposals for an action programme were drawn up in the late 1980s, but there had not been a large-scale breakthrough or generation of any momentum for an EU transport strategy until the advent of TENs in the 1990s (see Chapter 9).

One element of transport policy that did cause a lot of concern to local planning authorities in the mid-1980s was the issue of harmonisation of maximum permitted lorry-axle loadings. In the interests of removing non-tariff barriers to trade, an EU standard of 40 tonnes per axle was proposed. In countries such as the UK where this was heavier than previous maxima, there were fears for the impact on the environment and on the structure of roads and bridges of lorries of this size (ADC, 1985).

The contribution of the CoE

Work on a European regional or spatial planning framework had meanwhile been proceeding for several years under the auspices of the CoE. The CoE's Conference of Ministers responsible for spatial planning, CEMAT, had met at intervals since their first meeting in Bonn in 1970, when their declaration noted that European integration could aggravate geographical differences if not accompanied by a common approach to regional planning (EP, 1983: 20).

CEMAT established a working party to prepare a 'European Regional/Spatial Planning Strategy', under the direction of its steering committee on regional planning, CDAT. Involving as it did participants from CoE member-states in all parts of western Europe, not all of whom at that time anticipated EU membership, there was recognition that this strategy itself would not be any more than a basis for voluntary co-operation unless it was adopted on a binding basis by member-states or the EU.

What was achieved by this project was the development of the capacity to think spatially at the scale of the whole of Europe on the part of senior officials responsible in national governments for international aspects of planning policy, together with their professional advisers, and the articulation of the reasons why policy development on selected spatial planning issues could benefit from consideration at the supranational scale.

Some of the participants were from countries such as Sweden, which have since joined the EU. Significantly, in view of later events, much of the leadership and direction for this project came from the Dutch delegation and from the Rijksplanologische Dienst or National Physical Planning Agency. Luxembourg also provided an important resource in the form of the chief rapporteur who drafted the main report.

The title given to this CEMAT project, European Regional/Spatial Planning Strategy, is noteworthy in relation to the issue of terminology. Since the CoE operates in two languages, every effort is made to ensure that English and French documents say the same thing. The two words 'regional/spatial' were deemed necessary (and insisted upon by the UK government) in order to translate *aménagement* because the more appropriate of the two, spatial, was not then regarded as being sufficiently well understood by English-speaking planners. The word 'strategy' translates the French *schéma* quite legitimately, but this precedent is significantly not being followed in 1995 (see below).

Torremolinos Charter

Another outcome of the work of CEMAT at this time was the adoption by spatial planning ministers of the text of a *European Regional/Spatial Planning Charter* (CoE, 1984) (the same linguistic pedantry noted above applied to this). The charter aimed to define for the first time in a European context the notion of regional planning, its characteristics and objectives. Its protagonists saw it as a latter-day Charter of Athens.

The *European Regional/Spatial Planning Charter* became known as the Torremolinos Charter after the town in which the ministers met. This illustrates another cultural point. Those who want the charter to be taken seriously are not helped by the choice of city in which the ministers met to agree and sign it. To the Spanish hosts, Torremolinos is no doubt just another city which happens to have conference facilities, but in the minds of many British and other northern Europeans, Torremolinos is associated with package holidays, creating an impediment to hopes that it would be taken seriously by the planning community.

It was nevertheless a serious document, attempting to define the concept of regional/spatial planning and the role it can play in achieving balanced regional development and better spatial organisation in Europe, especially in response to problems which extend beyond national boundaries and underlie north–south and east–west differences. The fundamental objectives of spatial planning were set out as

- balanced socioeconomic development of the regions;
- improvement of the quality of life;
- responsible management of natural resources and protection of the environment; and
- rational use of land.

The charter sought to promote the idea that planning had a role to play at all spatial scales from local to European, that it should form part of the responsibilities of authorities at local, regional, national and European levels, that it should seek co-ordination between policy sectors, and that it should be democratic and participatory. A number of specific planning issues in urban, rural, regional, coastal, mountain and frontier areas were identified in order to demonstrate the need for planning policies. Although supportive of the CoE's own programme of work, the value of documents such as this is to encourage governments of countries that have not fully developed their planning systems to do so, and convince other European bodies, notably the EU, of the wisdom of doing so.

Iberian enlargement and the SEA

The accession of Spain and Portugal to the EU in 1986, closely followed by the passage of the Single European Act (SEA) in 1987, marked another step in the development of EU spatial policy. It also completed the Mediterranean phase of enlargement of the present EU, although applications for membership from Cyprus, Malta and Turkey are still pending. As with the second enlargement, the EU was admitting new member-states who were poorer than any of the existing member-states, thus creating new pressures on the funding programmes and EU budget.

Pinder (1983: 40) posed the question of whether Mediterranean enlargement would lead to 'increasingly strident demands for regional policy to deal with the problem of North/South disequilibrium'. The Integrated Mediterranean Programmes and subsequent cohesion policies (see below) bear witness to the validity of this prediction.

SEA

The SEA, which came into force on 1 July 1987, paved the way for the SEM programme by enabling all single-market measures to be adopted by QMV. It also amended the Treaty of Rome in a number of ways of interest in the spatial policy context. It added regional policy and environment policy to the formal competence of the EU, and Article 18 of the SEA amended the Treaty of Rome by adding several sectors of EU policy, including environment and the ERDF, to those on which decisions of the Council of Ministers would normally be based on QMV not unanimity.

The SEA added two new titles to the treaty basis for spatial policy, Title V on the Environment (Articles 130A to 130E), and Title XVI on Economic and Social Cohesion (Articles 130R to 130T). The term 'economic and social cohesion' needs to be given particular emphasis, as it has come to signify a central underlying theme and motive for spatial policy development, namely, the need to ensure that disparities of wealth and quality of life between the different regions of the EU are not allowed to widen to the point when they threaten the cohesion of the EU itself. This is sometimes left unstated, rather than made explicit, in individual policy sectors.

The SEM

The SEM programme, like the SEA, has its origins in a series of declarations at European summits in Copenhagen in 1982 and Brussels in 1985 calling for completion of the internal market. Following this agreement, a study was undertaken under the direction of Paolo Cecchini to examine in detail the steps necessary, in the form of removal of tariff and non-tariff barriers, to complete the internal market and to quantify as far as possible the economic benefits of the SEM and costs of 'non-Europe' if the barriers remained. This study has been published in book form (Cecchini, 1988).

The SEM programme itself was set out in the 1985 *White Paper on Completing the Internal Market* (Commission, 1985) from the Commissioner responsible, Lord Cockfield, which was endorsed by the Brussels Summit. This identified over 300 specific proposals or groups of proposals for legislation which would be presented to the Council of Ministers on a timetable geared to completion by

31 December 1992. In fact the final list was reduced to 282 measures. In spite of slippage, many were adopted well in advance of 1992 and the great majority were in place by the target date. A full list is given in Roberts *et al.* (1993). As the SEA stipulated that all internal market measures are subject to majority voting, these proposals could not be held up by national veto. Each measure had its own date for adoption and most were in the form of directives which had to be incorporated into national law. Some member-states, notably France, Greece, Italy and Portugal, were slow to do this (Owen and Dynes, 1993: 27).

The end-date of 1992, or in effect the beginning of 1993, did not therefore imply a big-bang introduction of the new internal market or SEM. Reference to the SEM as the 1992 programme was in effect a very successful presentational or public-relations device which grabbed the attention of politicians, the business community and the general public alike.

The Four Freedoms

A key concept of the SEM is the extension of the Four Freedoms throughout the EU. These are the free movement of goods, services, capital and labour. The SEM programme was based on a study undertaken for the Commission on the costs of 'non-Europe', i.e. an EU of internal borders where tariff and non-tariff barriers continued to apply. The argument was that these added greatly to business costs. Their removal would not only save money but it would also open up new business opportunities, release money for investment and stimulate economic growth and the integration of member-states' economies (Cecchini, 1988). The spatial adjustments which follow, or might follow, from this are a major factor in subsequent interest in the development of a Europe-wide spatial strategy (see below and Chapter 6).

The Four Freedoms affect planning in another way. Free movement of labour implies mutual recognition of professional titles, and one of the SEM directives addresses this issue. How this affects planning and similar professions is discussed in Chapter 13.

Co-ordination of the Structural Funds

A new Directorate-General, DG XXII, was set up in 1985 with responsibility for co-ordination of EU structural instruments. DG XXII undertook a thorough review of the funding programmes, proposing a framework regulation for the reform of the Structural Funds in 1987, leading to adoption of new regulations for the three Structural Funds by the Council of Ministers in February 1988 for the five-year operating period 1989–93. The three Structural Funds are the ERDF, the ESF and the guidance section of the EAGGF. From 1989, they have operated on a co-ordinated basis within a framework determined by five overall objectives, three of which, Objectives 1, 2 and 5(b), are explicitly spatial and apply to designated regions for which funding is made available under community support frameworks (for full definitions and description, see Chapter 7).

Although its terms of reference did not exactly correspond to those of the 'operational unit' called for by the EP on the basis of the Gendebien Report, and was by no means a direct response to this alone, it went some way in the direction that the Gendebien Report called for. It was also a response to the Commission's own assessment of the effectiveness of the funds and the extent to which they were still an aggregate of national government policies.

The February 1988 Council also resolved that the financial resources allocated to the Structural Funds should be doubled in real terms during the 1989–93 period, so that by 1993 they would account for one-quarter of the EU budget. Co-ordination of the Structural Funds and this increased budget allocation formed part of the so-called Delors I package tabled by the then President of the Commission, Jacques Delors, in 1987 (Scott, 1993a).

Community Initiatives and Innovative Measures

At the same time that the above arrangements were agreed, provision was made for an element of the Structural Funds to be made available to the Commission to deploy on its own initiative. This fell into two categories which together amounted to around 3% of the total Structural Funds budget. The categories were

- Community Initiatives (under Article 11 of Regulation EEC 4253/88); and
- Innovative Measures or Pilot Projects (under Article 10(1) of Regulation EEC 4254/88).

Community Initiatives are the successor to the CPs mentioned above, and have spawned an impressive array of acronyms. Several are of importance in the spatial policy context and are discussed in later chapters.

Innovative Measures or Pilot Projects (the terms are used interchangeably as the difference is really one of translation of texts in other languages), sometimes known as Article 10 measures, include a number of initiatives which are central to the theme of this book, including the *Europe 2000* and *Europe 2000+* studies (see Chapters 6 and 12), networking under the RECITE programme (see Chapter 8) and the Urban Pilot Projects (see Chapter 11).

IMPs

IMPs were a new spatial element in the EU's funding programmes, introduced in 1985 (*OJ* L 197; Commission, 1985) in preparation for the impact of enlargement. The political imperative was to assist the Mediterranean regions of France, Greece and Italy, which faced the prospect of increased competition within the EU for their products following accession of Spain and Portugal. Greece in particular pressed hard for a favourable financial arrangement and at one stage threatened to block the enlargement agreement in order to strengthen its negotiating position.

IMPs were significant in two ways for the development of EU spatial policy: its area of benefit made them a broad-brush spatial policy at the European scale, and they were an element in the move towards greater integration of EU funding programmes to achieve a more co-ordinated attack on the problems faced in specific regions or sectors of the economy.

The area of benefit included most of Greece, southern Italy and Mediterranean parts of France. Funding came from different sources and was not all new money. In addition to matching funds from local and national sources, in round figures 40% of IMP funding came from the existing EU funds, i.e. ERDF, ESF and EAGGF, 40% in the form of EIB loans and 20% was new money. This was to be spent on infrastructure and economic development projects such as the development of food-processing industries, agricultural diversification and irrigation projects, retraining or retirement of agricultural workers and rural development.

IMPs continued during the transitional period prior to full integration of Spain and Portugal, lasting until 1993, when the Cohesion Fund and the enlarged Structural Funds came into operation and took over their role.

Environment

The link between environment policy and balanced regional development was implicit in the SEA, a connection which the Commission sought to make explicit in the fourth action programme on the environment. It was later to be made explicit with the addition of the words 'balanced development of its regions' in Article 130r(3) in the TEU (see Chapter 10). During the 1980s, DG XI produced an ever-increasing flow of legislation (Haigh, 1989), much of it directed at specific pollutants. Concern for land use and spatial planning did not disappear but was expressed in green paper form (see below) rather than any legislative proposal similar to the environmental assessment directive.

The 1990s and Maastricht

The early 1990s saw two events of profound significance for Europe and the future direction of the EU, not least for their spatial policy significance: German reunification in October 1990 and the European summit at Maastricht in December 1991 which approved the Treaty on European Union.

In a way, these occurred in the wrong order. German reunification opened the way to a whole new set of priorities and future directions for the EU, while the Maastricht Treaty was in a sense the culmination of what had gone before. Both, however, were of vital significance to the development of EU spatial policy.

German reunification

The formal act of reunification of the German Federal Republic and the German Democratic Republic (GDR or DDR) took place on 3 October 1990, less than a year after the opening of the Berlin Wall in November 1989. Currency union had preceeded this in July 1990. As Chapter 2 points out, this was a classic case of responding to a new opportunity, as Monnet advocated.

This was in no sense a formal enlargement of the EU, since no new member-state was added and therefore there was no treaty of accession. The official doctrine was that one of the member-states was making an adjustment to its territory and boundaries. Reunification was thus regarded by the EU as an internal matter for one of its member-states. Six new federal states (*Länder*) were added to the German Federal Republic. In everyday speech, this territory is referred to as the new *Länder* (*neue Länder*), not as the former DDR.

This doctrine enabled the German federal government to proceed rapidly, without needing to seek agreement at a European summit and ratification by all member-state governments, as would have been the case if an enlargement treaty with the EU had been necessary.

Nevertheless, the addition of the territory of the former DDR, and the changes in central and eastern Europe that the fall of the Berlin Wall came to symbolise, set in train the most profound reorientation of the EU's thinking concerning its future enlargement, spatial structure and economy.

Some institutional changes did follow from reunification, reflecting the new political reality. There was no longer an approximate equality of size between the four largest member-states. Previously, France, Germany, Italy and the UK all had populations in the 50–60 million range. Now, Germany had a population of around 80 million. Allocation of seats in the EP was adjusted (see Chapter 3) to give 18 more to Germany and five extra to the other three largest member-states, France, Italy and the UK, although no other institutional changes in QMV or other arrangements in the Council and Commission were made.

Politically, regions of other member-states that were underdeveloped or in need of economic restructuring could no longer regard Germany as the rich paymaster of the EU without problem regions of its own, and argue that they faced worse problems. The task of restructuring the economy of the former DDR was enormous, exceeding many prior expectations. In spite of the doctrine concerning reunification referred to above, this was not a matter for Germany alone. Accommodation had to be found within the Structural Funds and other EU programmes, for example, for a substantial injection of EU resources. This had not been anticipated at the time of the 1988 reforms of the Structural Funds.

The consequences of reunification went beyond issues concerning the Structural Funds. The spatial implications were considerable both for Germany and for the EU, reflected in the spatial policy statements issued subsequently (see Chapter 6). In the Foreword to the *Guidelines for Regional Planning*, this is the first point the ministers make: 'Following the unification of Germany, the general setting for spatial development in our country changed fundamentally . . . At the same time, the growing degree of integration within the European Community as well as the radically altered situation in central and eastern Europe present regional policy with great challenges' (BMBau, 1993: I).

Incorporation of the territory, a former Comecon state, into the EU confronted all the EU's policy-making structures with the reality of the environmental and economic legacy of the centrally planned economies. At the same time, several other newly westernised central European countries saw that the DDR had moved with great rapidity all the way from being a Comecon economy to full membership of the EU, and wanted to follow as quickly as possible.

After a phase of Mediterranean enlargement, followed by northern enlargement, the political imperative to enter into a phase of eastward enlargement of the EU has become unavoidable. Existing thinking and policy within the EU, and indeed statistics and the knowledge base for policy-making, has for so long been bounded by the EU of 12 that it has taken time for the change in direction to be reflected in its aspects. The broader spatial implications of reunification and the prospect of eastward enlargement, for example, were not reflected in the *Europe 2000* and *Perspectives in Europe* reports published in 1991 (Commission, 1991; RPD, 1991) but were in *Europe 2000+* (Commission, 1994a) and other later policy developments (see Chapters 6, 8, 9 and 12).

In fact, preoccupation with the problems of transformation of CEE countries has become a leitmotiv of many EU spatial programmes. Either they are specifically designed to allow for the future integration of CEE (for example, the TENs), or they are anticipating the lower priority which the poorer regions of the south and west can expect in an EU of over 20 countries (for example, the Cohesion Fund).

Spatial planning in the Maastricht Treaty

In the two to three years prior to the Maastricht Summit in December 1991 there was a noticeable quickening of the pace in respect of spatial policy thinking. The French presidency called the first of what was to become a regular series of informal meetings of ministers responsible for spatial planning in Nantes in 1989. As a result, the Commission was invited to prepare a document setting out the EU's approach to spatial planning and the context for EU spatial policy. The outcome was the *Europe 2000* report prepared by DG XVI entitled *Outlook for the Development of the Community's Territory* (Commission, 1991).

Meanwhile the Environment Directorate, DG XI, under one of the most energetic commissioners ever to hold the environment portfolio, the Italian Carlo Ripa de Meana, undertook the task of preparing proposals for the development of an urban policy dimension to EU environment policy. The outcome of their deliberations was published in 1990 as a Commission green paper, the *Green Book on the Urban Environment* (Commission, 1990b; see Chapters 10 and 11). This was published in 1990 and launched with wide publicity at a conference in Madrid in April 1991 presided over by the British Prince of Wales. However, the green book proved to be something of a false dawn for EU urban policy.

A number of studies by national governments were undertaken at this time which also addressed the concept of supranational spatial planning at least to the extent necessary to demonstrate the need to set national policy within such a context. The most important of these from the point of view of EU policy development was the *Perspectives in Europe* report prepared by the Dutch National Physical Planning Agency, the Rijksplanologische Dienst and published shortly before the Maastricht Summit (RPD, 1991; see Chapter 6).

Informal councils

The meeting of ministers in Nantes in 1989 was the first of what has become a regular series of meetings. Since the EU does not have a formal competence in spatial planning, these meetings had the official status of informal meetings. After Nantes in 1989, later meetings were in Torino in 1990 and Den Haag in November 1991, shortly before the Maastricht Summit. Being informal, at first not every presidency felt the need to hold such a meeting but they have generated a momentum and work programme that has established them as a regular feature of later presidencies. At the meeting in Liège in 1993, it was decided that they should have the status of informal councils. Each presidency since has called such an informal council. Although not able to adopt EU legislation as a formal council could, they have proved to be a significant driving force in the formulation of EU spatial policy.

The German presidency put considerable effort into preparations for the Leipzig meeting in September 1994, and both this meeting and that of the French presidency in Strasbourg in March 1995 took important steps to develop EU spatial policy further, a process continued at the meeting in Madrid in November 1995 (see Chapters 12 and 15).

Maastricht outcome

The negotiations surrounding the Maastricht Treaty, just as much as the Treaty on European Union itself, are of considerable interest to spatial planners. Both the Dutch government and some members of the Commission, with the support of some other member-state governments, wanted to extend the scope of EU spatial policy by amending the treaty to create an EU competence in spatial planning. Others were arguing that a competence in urban policy should be added. The motivation for studies such as *Europe 2000, Perspectives in Europe* (see Chapter 6) and the *Green Book on the Urban Environment* (see Chapter 10) was partly explained by this factor. Their ambitions were not fully satisfied by the final text, and will therefore figure in the discussion at the 1996 IGC (see Chapter 14).

Nevertheless, a lot was achieved by the Maastricht negotiations. In the first place, and somewhat to the surprise of many observers, the treaty includes reference to 'town and country planning' in the environment title, Article 130s. While in its British sense this does not equate with spatial policy, the equivalent text in the French and German versions comes very close to doing so (see Chapter 4, Figure 4.1). This provision is however subject to unanimous voting. It is more of symbolic significance (Williams, 1993a) and not likely to have much impact unless it is changed to QMV, and nothing has been adopted under it at the time of writing.

Of much greater importance was the increased emphasis on economic and social cohesion – the creation of the Cohesion Fund to assist the four poorest member-states, Greece, Ireland, Portugal and Spain (see Chapter 7), the new title on TENs which has reinvigorated EU transport policy (see Chapter 9), and enhanced environment powers (see Chapter 10). The creation of the CoR (Chapter 3), although representing subnational government rather than specific regions, is likely to increase pressure on the Commission and Council to address spatial issues and has clearly identified EU spatial policy as one of its priority concerns.

CSD

At a more mundane level, new working arrangements for spatial policy development were agreed at this time in the form of the CSD (see Chapter 3). This is an intergovernmental committee of senior officials from national governments, who meet to co-ordinate spatial policy development and have the task of preparing policy guidelines in the form of a *European Spatial Development Perspective* (*ESDP*; see Chapter 12).

Environment

Environmental politics was centre stage in the period following the 1989 elections to the EP, which elected sufficient Green Party MEPs for them to form a new party group, known as Les Verts. Thirty were elected from seven countries, Belgium, France, Germany, Italy, The Netherlands, Portugal and Spain. Although none were from the UK owing to the workings of the British electoral system, the UK did record the biggest percentage Green Party vote.

The Commissioner then responsible, Carlo Ripa di Meana, ensured that it had a high profile in the Commission's policy-making. It was in this climate that the Maastricht negotiations took place in 1991.

As Chapter 10 explains, the treaty requires that all EU policies and actions must be based on the highest environmental standards. However, the Environment DG did not succeed in one of its objectives, to have a competence over the urban environment added to the treaty. The inclusion of 'town and country planning' fell far short of its ambitions in this respect. This, plus the departure from Brussels of the energetic environment Commissioner Ripa de Meana, meant that some of the impetus went from EU policy-making on the environment in 1992–4.

The pendulum will swing back again. With the European Environment Agency becoming operational in 1994 and accession of Austria, Finland and Sweden in 1995, the balance of power between the more and less environmentalist countries will shift substantially in favour of those seeking to adopt higher minimum standards. Given the treaty powers and the much wider applicability of majority voting than before the Treaty of European Union was adopted, the scope for any member-states wishing to block new environment legislation will be much reduced.

Funding

The budget for the reformed Structural Funds doubled in real terms in the period 1987–92 under the Delors I plan, increasing from 17% to 27% of the total EU budget. On the basis of the Delors II package, presented in the white paper, *From the Single Act to Maastricht and Beyond: The Means to Match our Ambitions* (Commission, 1992a), the Edinburgh Summit in December 1992 agreed the allocation for the Cohesion Fund and another doubling of the size of the Structural Funds over the period 1994–9. The allocation for 'economic and social cohesion', of which the Structural Funds was the main item, was projected to reach 33.5% of the entire EU budget by 1997 (Scott, 1993a: 75). The essential features of programme funding based on five objectives, and an emphasis on partnership, subsidiarity and additionality remained as established in the 1988 review, although the bias towards Objective-1 regions introduced in 1988 was strengthened and greater discretion to adjust Objective-2 boundaries to match local conditions was agreed (see Chapter 7).

White paper on Growth, Competitiveness and Employment

This white paper, the full title of which is *White Paper on Growth, Competitiveness and Employment – the Challenges and Ways Forward into the 21st Century* (Commission, 1993b), is so closely associated with the then president of the Commission that it is commonly referred to as the Delors White Paper. It was adopted by the Brussels Summit in December 1993 and is a reference point for a whole range of policy development since that date.

It was prepared as a response to the growing rates of unemployment and shrinking economic growth rates throughout Europe in the early 1990s, but a more fundamental political purpose on the part of Jacques Delors was to try to regain the initiative and restore some momentum for the EU after the divisive debates over ratification of the TEU.

The Delors White Paper seeks to establish a framework for future policy-making designed to ensure lasting and sustainable economic growth, and recommended that the EU sets itself the objective of achieving an annual growth rate of 3% p.a. and creating 15 million new jobs by the end of the century. The white paper did not itself propose new programmes of expenditure, new EU legislation or powers. Instead, it sought to guide policy development in sectors where EU competence already existed. TENs, regional policy, education and training were prominent among those identified. In response to the political realities of the time, it also paid great respect to the principle of subsidiarity. It clearly differentiated between action by the EU and action at the national and subnational levels of government. Instead of prescribing to the member-states, it put forward a menu of possible actions from which governments could select those deemed most suitable to their own situation.

The white paper has three parts. Part A summarises the Commission's vision and strategy for the achievement of a thriving, competitive and decentralised economy with high employment levels for the EU as a whole, to ensure economic and social cohesion and sustainable growth. Part B is the core of the policy programme, while Part C summarises the contributions by the member-states.

Among the priorities for action are the development of TENs for transport, telecommunications and energy in order to improve competitiveness within the internal market and overcome problems of peripherality. Linked to this is the promotion of information technology and encouragement of SMEs. The white paper also suggests that new approaches to urban planning and policies to reduce energy consumption should be developed.

Transport and TENs

After having been, since the original Treaty of Rome, the sector conspicuous for its lack of progress towards EU policies, new life has been breathed into this sector by the Maastricht Treaty. A key feature of the treaty is the new Title

XII on TENs, intended to bind the EU together both physically and metaphorically. For the first time, in the 1990s there is the prospect that a spatial policy framework for transport infrastructure at the EU scale will be developed.

The fourth enlargement in 1995

The enlargement that took place on 1 January 1995 was officially the fourth enlargement, although German reunification had in effect enlarged the EU, as well as introducing unprecedented economic and environmental problems. Not only did it extend the EU northwards but, like German reunification, it also extended it significantly eastwards.

Clearly, the fourth enlargement was of great significance in relation to the changing spatial structure of the EU as a whole, adding as it did approaching 40% to the surface area of the EU. For the first time, apart from the period 1973–84 when Greenland was still part of Danish territory and therefore part of the EU, a significant part of the EU's inhabited territory was within the Arctic.

Unlike the 1981 and 1986 enlargements, the new member-states were prosperous and did not notably extend the economic disparities within the EU. There was less reason therefore to introduce new financial instruments such as the IMP or the Cohesion Fund. Nevertheless, spatial issues figured in the enlargement negotiations. Arctic agriculture and the support of Arctic communities in the face of an open market in agricultural produce was a new issue for the EU.

The three Nordic applicants (Norway played a full part in these negotiations – the referendum in which its voters rejected EU membership was only five weeks before the proposed accession date) argued that special measures would be necessary for their Arctic regions and sought inclusion of appropriate areas within Objective 1 or at least Objective 2 of the Structural Funds (see Chapter 8). None of the areas concerned had low enough standards of living to meet the per capita GDP criteria for this, so the solution found was to create a new Objective 6. This is in effect an Arctic version of Objective 1 for areas with a population density below eight persons per square kilometre. It exists only in Finland and Sweden, its legal basis being the accession treaty. Secondly, of particular importance to Austria was the question of transit routes across the Alpine passes.

Thirdly, the question of second homes and of associated land-use planning and property law was an issue of sufficient significance to play a part in the enlargement negotiations as such, although some of the negotiators from the existing EU of 12 had assumed that these could be left to the local and regional planning authorities of the countries concerned.

The fourth enlargement is regarded as having possibly been the last orthodox enlargement, in which the acceding states can be expected to accept the *acquis communautaire* within a reasonable timescale. The next phase is likely to involve the former Communist Party states of central Europe which will, like the former German Democratic Republic, pose great economic and environmental strains. Up to ten such countries are in the queue for membership, along with Malta and Cyprus. Enlargements in the next decade could make further dramatic changes to the spatial structure and scale of Europe.

6

Ways of looking at EU space

Having set out in Chapter 5 the different policies and programmes that now form elements of EU spatial policy as it has evolved up to the mid-1990s, this chapter turns to the question of 'thinking European' in spatial planning. Just as a local planner needs to develop the mental capacity to grasp the spatial scale and inter-relationships that exist, or potentially may exist, within a local plan area, so at the European scale the same ability needs to be developed. The difference is that the European scale is outside the professional experience of the majority of planners, and those from some of the larger countries many find it difficult to think supranationally for reasons of education and upbringing as well as lack of direct experience.

Several ways have been attempted to help overcome this problem, some more abstract than others, and there have been several quite substantial spatial policy-making studies during the 1990s which aim as much to aid the supranational thinking process as to prescribe policy as such. This chapter focuses on supranational spatial thinking, and Chapter 12 on the state of policy prescription.

An additional complication at the European scale is the changing spatial structure of the EU itself, as is immediately apparent from consideration of successive enlargements. There are other factors as well which can add to the complexity of switching from local or regional to supranational planning. This chapter outlines these first, then considers the more abstract theme of conceptual models, metaphors, doctrine and ways of thinking that are applied to Europe's space. It then goes on to look at a number of approaches to supranational planning that have been undertaken by both the EU and member-states.

Changing spatial structure of Europe

The EU is subject to changes of its spatial structure to an extent not normally experienced by individual nation-states, and not solely the result of the successive enlargements. Attention is turned now to those brought about by other factors. They fall into two broad categories: infrastructure development and locational decisions. Their implications need to be grasped if the capacity to think European is to be acquired.

The Channel Tunnel between England and France is a prime example of the first category. It is infrastructure of profound spatial significance, bringing to an end Britain's island status with its opening in 1994. Perhaps the profundity of this change and the challenge it poses to the island mentality explains the contrast between the UK and France in the way in which the respective countries showed an ability to think strategically about their spatial position in Europe and the opportunities offered by the tunnel.

Fixed links among Sweden, Denmark and Germany will have a substantial impact on the spatial structure and relationships of that part of Europe, and will also require considerable readjustment in the thinking of planners in the islands of Denmark or in southern Sweden. The *Denmark 2018* report (see below) represents the first stage in pushing this readjustment along.

The bridge and tunnel across the Storebaelt (Great Belt), connecting Zealand with Funen and the Danish mainland, due for completion in 1995, is under construction. The rail link is due to come into operation in 1997, and the road link early in 1998. After some delays owing to environmental objections, the Oresund bridge and tunnel link between Malmö in Sweden and an artificial island near Copenhagen airport is going ahead with an anticipated completion date of 2000 (see Chapter 8). This will accommodate a motorway of two lanes in each direction plus railway tracks. When these are operational, Sweden will be linked directly for the first time to the mainland of western Europe.

If another proposal goes ahead, to build a fixed link from the Danish island of Lolland, south of Copenhagen, to Germany across the Fehmarn Belt, then western Sweden will become very accessible to northern Germany, Hamburg and Berlin, while parts of Denmark that at present carry a lot of through traffic will in effect be bypassed. Another spatial effect of these proposals will be to increase the relative peripherality of Finland. At present, both Sweden and Finland are cut off from the European mainland by sea crossings. Sweden's spatial situation within the EU would be greatly enhanced by completion of the fixed link to Denmark, increasing Finland's disadvantage relative to Sweden.

The network of inland waterways are an important element in shaping the spatial relationships of the west European mainland. Construction of a Main–Danube canal link in southern Germany has been an ambition since Charlemagne first attempted the project in 793. Bavaria's King Ludwig I did actually succeed in completing a narrow-boat canal in 1845 but in the twentieth century this became little more than a tourist attraction.

Completion of the link capable of taking normal-sized Rhine barges across the 170-km stretch of land between Bamberg on the Main and Kelheim on the Danube in September 1992 may not have the same short-term impact as the other fixed links described above, but in the longer term it may also facilitate substantial changes in spatial relationships. It offers the opportunity for waterborne trade with Germany and The Netherlands, and access to Rotterdam, and cities such as Vienna, Bratislava and Budapest.

The construction of north–south rail tunnels through the Alps in Switzerland is another infrastructure development of European spatial significance. A referendum in September 1992 resulted in a 'yes' vote for this proposal, which is due for completion by 2010. In spite of this vote, the issue is still controversial and delays may occur as more detailed plans are proposed. At the time, the 'yes' vote was interpreted as an indication that the Swiss electorate may be more receptive to proposals to integrate Switzerland further in the EEA and eventually the EU. However, a subsequent referendum on 6 December 1992 rejected the EEA agreement (see Chapter 3).

As a non-EU country, Switzerland has been able to impose its own restrictions on transit traffic, creating two distinct overland flows in and out of Italy, via France and via Austria. Transit traffic is a particularly sensitive issue in Austria, especially in Tirol in the west, and was a key element in enlargement negotiations.

The locational decision which may turn out to be of the greatest long-term significance for Europe's spatial structure is in the first place a national issue. This was the decision of the German Bundestag on 20 June 1991 to designate Berlin as the federal capital of reunited Germany. Whereas Bonn is situated within the Blue Banana (see below) or core region of the EU, easily accessible to Brussels and other capitals, Berlin is well outside this core region. However, as the capital of Germany, Berlin will be of sufficient European significance to act as a counterweight to the Blue Banana and generate its own network of links, communications and development pressures. Other locations currently peripheral to the core of the EU but which have good access to Berlin will cease to be peripheral in European terms. Poland and the Czech Republic will benefit from this effect, while Denmark and Sweden rather than France and the Benelux countries will become the nearest 'western' neighbours. The infrastructure development referred to above is clearly not unrelated.

Another way in which the spatial structure of the EU changes, or is perceived to change, is as a result of locational decisions by the EU in respect of its own institutions. Whereas most nations have one acknowledged national or federal capital and clearly established practice for deciding the location of national institutions, this is not the case with the EU. Brussels exhibits many of the characteristics of a capital, notably as the seat of administration (the Commission) and of key decision-making (the Council). The scale of the lobbying industry's presence (see Chapter 8) certainly reinforces the view that Brussels could be regarded, *de facto*, as the capital, but this has no formal status. Nor are all the institutions of a type normally associated with capital-city status located there since the ECB and ECJ are elsewhere and the EP is peripatetic.

The location of the main EU institutions is settled as set out in Figure 3.5, Chapter 3. Two of the most important, however, the Parliament and the Commission, still have provisional locations. The most irrational, if one disregards national pride and politics, is that of the Parliament.

Among locational decisions for EU institutions taken in the 1990s, that of the EMI and the proposed ECB were regarded as the big prizes. London, seeing itself as Europe's *première* financial centre, believed that it had a natural claim to the ECB, but the British opt-out on the single currency, negotiated as part of the Maastricht Treaty, meant that this claim could not be taken seriously in other capitals. Furthermore, selection as the location of another EU financial institution, the EBRD, had the effect of relinquishing rather than reinforcing London's claim. The choice of Germany's financial centre, Frankfurt am Main, for the ECB is expected in due course to promote the ambitions of Frankfurt to become a financial centre of worldwide and not just European significance.

The rotating presidency means that at present every member-state, however peripheral its location, can look forward to a period when it plays a central role in EU policy direction and hosts a European Council and other key meetings. If arrangements for the presidency are changed at the IGC into some joint or shared arrangement, this periodic centrality may no longer apply. Member-states which feel they may become marginalised as a result may be expected to look even more closely at future locational decisions of EU institutions.

Peripherality and ultraperipherality

Preoccupation with the core of Europe and the location of the main institutions is matched by preoccupation with peripherality. This concern lies at the heart of policies for economic and social cohesion, including the Structural and Cohesion Funds, and is the subject of many critiques of the outcome of EU spatial policy instruments and of concerns about greater economic integration. Normally, these concerns are expressed on behalf of countries such as Portugal and Ireland, or industrial regions such as those of northern Britain.

It should not be forgotten that the territory of the EU is not entirely located in Europe. The overseas territories of several member-states are or have been included as part of that country's territory for the purposes of EU membership.

Of the original six member-states, France included its *départements outremer* (DOMs), so from its beginning the EU contained territory on the South American mainland (Guyane), Caribbean (Guadeloupe and Martinique), Réunion in the Indian Ocean and islands of St Pierre et Miquelon off the shore of Newfoundland in Canada. Subsequent enlargements brought Greenland into the EU in 1973 until its independence from Denmark in 1984 and, in 1986, Ceuta and Melilla on the north African mainland, and the Canary Islands (Spain), the Azores and Madeira (Portugal) in the Atlantic.

The UK is an exception. Not only did it not bring any remote territories into the EU but also three constituent parts of the British Isles that are clearly European and not especially peripheral, Jersey, Guernsey and the Isle of Man, are not members of the EU, and neither is Gibraltar – although it does benefit from some programmes. If constitutional arrangements similar to some other member-states had been adopted by the UK, not only these but perhaps also the Falkland Islands or Bermuda would have had the status of EU membership.

These remote parts of the EU are not to be viewed simply as curiosities. They are entitled to participate fully in funding and other programmes, and their conditions and requirements must be taken into account when structural and spatial policies are formulated and implemented. They can elect members of the EP and one of the 24 French members of CoR comes from Guadeloupe.

Spatial metaphors

It is worth referring here also to the proliferation of spatial metaphors being generated in order to help us to think spatially in relation to Europe as a whole. In their analysis of the very successful 'Green Heart' metaphor that is central to Dutch spatial strategy and to the way many Dutch citizens think of their national territory, Faludi and van der Valk (1994: 67) argue that 'metaphor can be a pervasive mode of understanding by which we use experiences in one domain to structure another . . . It conveys meaning and intent'. Also in the context of spatial policy, Shetter argues that 'Metaphor has a critical role in human knowledge and action and is central in human imagination, providing a quasi-logical framework of associations' (Faludi and van der Valk, 1994: Shetter, 1993: 67).

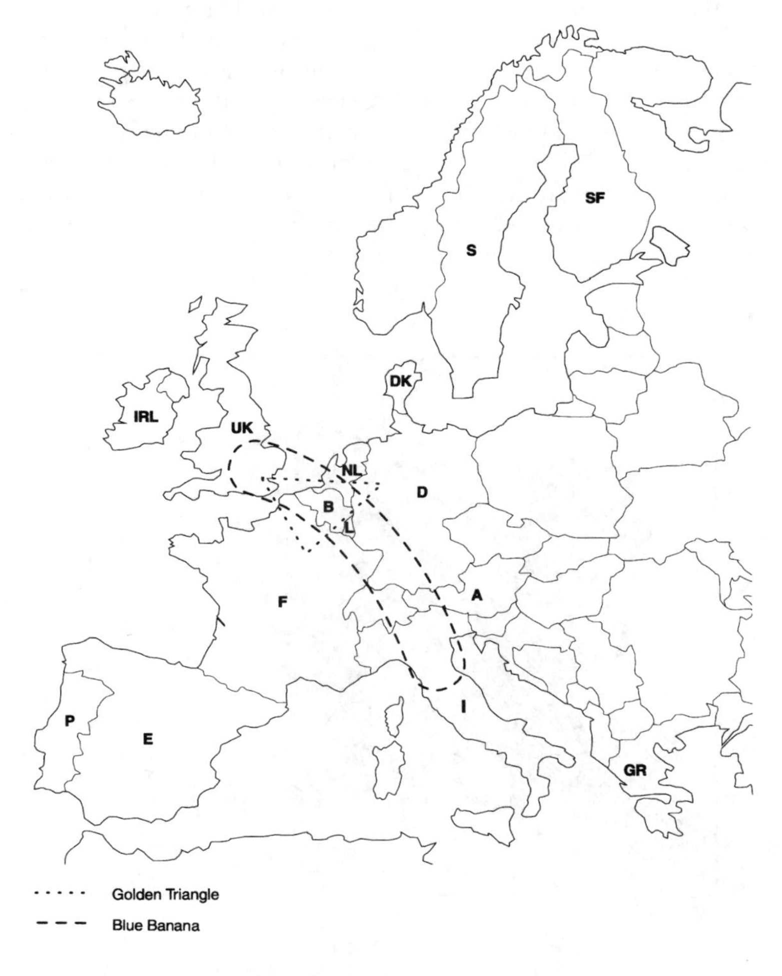

Figure 6.1 *The Golden Triangle and the Blue Banana*

The metaphor which started it is that of the Blue Banana (Brunet, 1989), referring to the economic and political core of Europe, running roughly in the shape of a banana from Milan through Zurich, Frankfurt, Bonn, Brussels and Amsterdam to London. This metaphor creates a memorable image which simplifies and structures people's thinking about the spatial structure of Europe (Figure 6.1).

The image of the Blue Banana is very much based on a core-periphery construct, which is very familiar to many of those responsible for issues of regional policy and government in both the UK and France, and is found also in the traditional idea often taught in school geography lessons of the golden triangle of London–Paris–Ruhrgebiet (Williams, 1993).

An alternative, put forward on the basis of experience of postwar Federal (former west) Germany, is the European bunch-of-grapes metaphor (Kunzmann and Wegener, 1991; Figure 6.2), which conveys a polycentric image of Europe's urban and economic structure. This is surely more appropriate at the continental scale, and also allows for the possibility of gaps of underdevelopment between the development centres. This issue is now reflected in the designation of areas of benefit from the Structural Funds (see Chapter 7), which includes several that are quite central in overall European terms. Japanese corridors, alluding to the shape of a Japanese *kanji*, is another formulation which expresses a dispersed spatial structure, this time around linear axes of development and movement (Kunzmann, 1992).

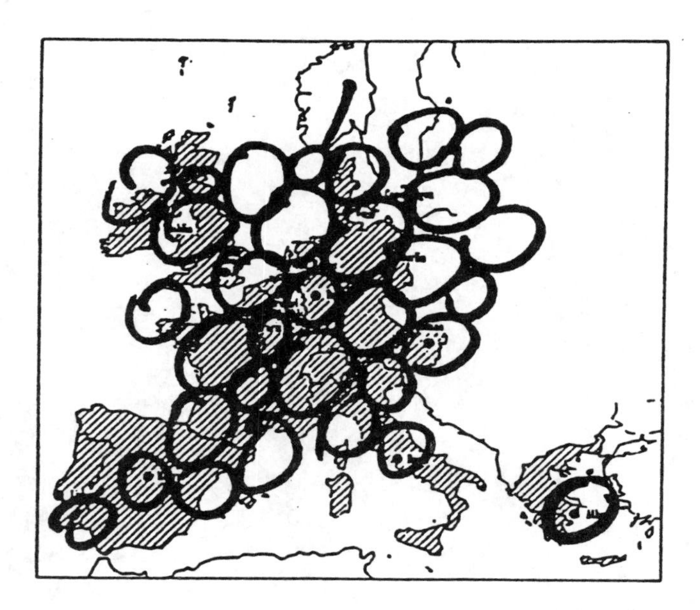

Figure 6.2 *The European Grape* (*Source*: Kunzmann, 1992).

Metaphors such as these are not predictions, and certainly not predestination, although people sometimes talk in these terms. This is especially true of the Blue Banana, which is sometimes taken to imply that all other areas are doomed. They really should be regarded simply as ways of describing the spatial structure of Europe in a manner that can be easily grasped, which may help people who find it difficult to think in European spatial terms to gain a sense of spatial positioning, and may also help with place marketing.

Spatial metaphors such as the Blue Banana may also be referred to as if they could form the basis of a planning doctrine, in the sense of a consistent theme underlying successive spatial strategies. The concept of a planning doctrine, and the importance of the Green Heart metaphor in providing a consistent theme in the development of Dutch planning thought, has been so clearly shown by Faludi and van der Valk (1994). The new strategic plan under preparation for Flanders is following the Dutch example in seeking a strong spatial metaphor, that of the Vlaamse Ruit or Flemish diamond for the core urban network of Gent–Brussels–Leuven–Antwerp, in order to create an image and concept with which to face up to EU competition, assert the identity of a Flemish core region, and concentrate the minds of local planners and politicians on the need for spatial planning policies. The question is raised by these approaches of whether there could in the future be a European planning doctrine.

Spatial positioning

Most local planners have a clear sense of the location within national space of the place for which they are responsible, often without thinking very consciously about it. The capacity to conceptualise or think about one's location or situation within the spatial structure of Europe as a whole is a skill which often needs to be developed. Spatial positioning is the term proposed for this skill. Through such a process, it is sometimes possible to identify opportunities, comparative advantage and possibilities on the basis of which new links and relationships could be developed and strategic policies formulated. Kunzmann (1992) has drawn attention to the process of spatial specialisation, for example in financial services, research and high technology or cultural industries. This is a good example of the directions such an approach may lead to. Spatial positioning has been a particular characteristic of French policy development, and the French strategy for the development of Europoles (DATAR, 1990) is a good example of its application. It is also a feature of many city-imaging and promotional activities, and of activities outlined in Chapter 8. In many ways, it requires imagination and lateral thinking rather than any particular technical skill.

An application of the techniques of discourse analysis to marketing and promotional material from two business parks, one in the Savoie near Grenoble, France, and one near Sunderland in north-east England (Peizerat, 1995), reveals these different approaches very clearly. The French example is promoted on the basis of its location in European space, whereas the British example stresses the locational features of its immediate surroundings and its international style designed to appeal to far-eastern and North American business. Location in the EU is seen as a political and economic given rather than a spatial factor offering advantages.

The story of the Channel Tunnel rail link vividly illustrates the difference between the more conceptual and pragmatic modes of thinking as applied to a unique project rather than to projects facing strong competition from rival developers. On the French side, its spatial relationship to the TGV system was thought through in a way that led to a national strategy for transport infrastructure (Holliday *et al.*, 1991), while the UK was totally preoccupied with the issue of the rail link to London and exhibited a total lack of thinking at the national spatial scale. Holliday *et al.* identify the French way of thinking associated with *aménagement du territoire*, and its corresponding absence in Britain, to explain this.

With more imaginative approach, and a sense of positioning within the EU policy-makers could have envisaged, for instance, Liverpool as the Atlantic port of Europe with direct high-speed rail links to the tunnel, so that it could compete effectively with Rotterdam just as it once did with Hamburg. Similarly, a direct line to Southampton or Bristol from the tunnel could have played a role in spreading the benefit of the tunnel nationally while meeting the aims of those concerned about overcongestion and unacceptable development pressures in Kent (Williams, 1991).

This leads on to a second way in which spatial positioning at the EU scale may help to overcome the mental blocks of orthodoxy. In most countries where there is a dominant capital-city region, such as France and the UK, it is very easy for those in smaller regional centres to accept subsidiary status, and to imagine that communication with the rest of Europe must be through the national capital, either metaphorically or literally, because of the transport infrastructure or hierarchy of government responsibility. In several countries, a sense of bypassing the national capital has an obvious appeal to many of those concerned to promote the European profile of a regional city. This is evident from the amount of lobbying and networking that such cities undertake, and from the growing numbers of local and regional representative offices in Brussels (see Chapter 8). It is also evident from the growth and loading of many airline services which allow travellers from regional cities to use an airport in another country as their hub airport.

The Northern Arc

An excellent example of the application of a form of spatial positioning to the formulation of a strategy for the promotion of regional interests by developing east–west trade and transport links is that of the Northern Arc. The concept is of an east–west corridor stretching from northern parts of Ireland through the north of England to Denmark, northern Germany, Poland, Scandinavia and all the Baltic states to St Petersburg. The objectives are to identify opportunities for trade and tourism, weaknesses in transport and communication infrastructure, and issues relating to rail, air and shipping policy; to promote mutual self-help for peripheral regions of Europe and environmentally friendly transport;

to motivate other sectors of the economy and agencies of government; and to create a TEN of interdependent regions. Above all, it is a concept aimed at bypassing the congestion and the decision-making power of the capital cities in the Blue Banana.

The Northern Arc concept identifies the Baltic Sea region as an example of a potential new core of development, one recognised by the *via Baltica* concept. This is an example of a rhetorical concept or vision of a future transport route linking St Petersburg with Tallinn, Riga, Vilnius, Warsaw and Berlin, which may in due course find practical expression through the development of TENs (see Chapter 9) and the *Vision and Strategies around the Baltic Sea 2010*, discussed below.

Supranational planning studies

As outlined in Chapter 2, the CoE had pioneered the concept of a European spatial development strategy during the 1980s. The outcome of this work was not so much a plan within which national and regional strategies were formulated as the means by which a number of issues of common interest to several countries, or of significance beyond the borders of an individual country, were highlighted. Perhaps the most important aspect of this work programme was the extent to which it was an educative process for the senior government officials and spatial planning experts. It enabled them to see the possibilities for this scale of planning and they were therefore both more receptive and better equipped to meet the challenge of supranational planning within the EU.

Europe 2000

The *Europe 2000* report, *Outlook for the Development of the Community's Territory* (Commission, 1991), was the first attempt by the Commission to provide the spatial planning community with a reference framework at the EU rather than national or regional scale for their actions. In a sense, it was the type of study the EP called for in 1983 when it adopted the Gendebien Report (see Chapter 5).

As the then Commissioner, Bruce Millan, justifiably stated in the Foreword, 'Europe 2000 breaks new ground in considering regional planning at the European level'. He went on to commend some of the studies being undertaken by national governments, such as those of Denmark and The Netherlands discussed below, and of local and regional authorities that recognise the European spatial dimension in preparing their own strategic plans. An example is the 'Barcelona 2000' study, which set the city in the context not just of Catalonia but also of the rapidly developing Mediterranean littoral.

The *Europe 2000* study itself was funded under Article 10 innovative measures of the ERDF regulation (see Chapter 7). It was undertaken at the invitation of the first informal meeting of regional planning ministers at Nantes in 1989, and the final report, officially a communication from the Commission to the Council, was presented to the third meeting of ministers at Den Haag in November 1991, shortly before the Maastricht Summit.

Although *Europe 2000* was in no sense a spatial plan for the EU as a whole, and in fact did not move as far in the direction of policy prescription as some people hoped, it was clearly intended to represent the start of a continuing process of developing greater coherence in EU spatial policy. It was designed to stimulate debate and sharing of ideas, information and experience, and undoubtedly did capture the attention of the spatial planning community and local and regional levels rather than be dismissed as being solely of concern to those few planners working at the national and supranational scales.

Much of the contents of *Europe 2000* consist of a geographical analysis of the EU, illustrated with detailed case studies or analyses of specific issues and examples of EU policies. It does not confine itself to policies within the remit of DG XVI although the team that prepared it was from that DG and the Regional Policy Commissioner took political responsibility for it.

Following a review of the existing state of EU policy development and of initiatives by national governments and individual regions and cities which take a European view, the introduction sets out the reasons why the Commission felt that greater policy coherence was needed. The first section sets the demographic and economic context, identifying the present pattern of regional development in the EU and the factors affecting location and the changes taking place in production and location patterns.

The second part, on infrastructure and spatial coherence, is concerned with transport, information technology and energy issues. Transport is seen as a sector in which problems are intensifying as the SEM is completed, new trading relationships develop and links with EFTA and CEE countries grow in importance. Links to peripheral regions are inadequate while congestion is endemic in certain core locations. Part of the problem lies in the extent to which transport networks have been developed within national boundaries throughout Europe, although the need in the 1990s is to meet European rather than national demand. Capacity and use of existing transport infrastructure, ports and airports, as well as road, rail and inland waterways, are analysed. New infrastructure that is under construction or proposed is reviewed in order to ascertain the extent to which it is meeting the needs identified at the EU scale, and suggestions are made for EU action to complete missing links in the transport network, several of which correspond to priority proposals in the TENs programme (see Chapter 9). One of the case studies included in this section is on the anticipated regional impact of the Channel Tunnel.

The development of TENs and of connections between different modes of transport is seen as the key to overcoming peripherality, as is exploration of the extent to which information technology and telecommunications (ITT) can substitute for travel. A key question posed by the discussion is whether wider adoption of ITT will increase or reduce regional disparities, whether it will spread decision-making, research and innovation throughout Europe or whether it will allow firms to concentrate top decision-making in fewer offices located in the core of the EU. The fear is that the most innovative technology and highest capacity will be installed in areas that are already the most highly developed, reinforcing their advantage.

The report draws attention to Community Initiatives (see Chapter 7) under the Structural Funds which were then in operation to counteract this. These included STAR (Special Telecommunication Action for Regions), which was concerned with investment in infrastructure, and TELEMATIQUE, aimed at improving telematic services in less favoured regions.

The final part of this section considers energy supply and the existing levels of integration and capacity of supply networks. Key problems inhibiting economic and regional development are identified, especially in relation to Greece, Spain, Portugal, the former DDR and crosswater links to the UK and the Nordic countries, and the case for energy planning as a component of an overall EU spatial strategy is argued.

The third section is on environmental issues. It starts from the propositions that economic growth and environmental protection need not be contradictory objectives, and that economic activities can be adjusted to avoid further degradation and promote sustainability. The main sectors of environmental concern are reviewed, and spatial patterns of air pollution, water resources, waste disposal and natural heritage conservation examined in order to assess their impact, and the potential role of spatial and land-use measures in overcoming the problems.

The fourth section focuses attention on specific types of area and spatial phenomena to which EU policy is or may be directed: urban areas, rural areas, border cities and regions, coastal areas and islands. Changes in their spatial structure and relationships, and on planning issues faced in these areas, are analysed, and existing EU programmes reviewed. The growth of co-operation and networking between cities and regions is particularly noted.

Throughout *Europe 2000*, policy implications and issues which could be addressed within an EU spatial framework are noted, and these are drawn together in a relatively short final section. This does not aim to set out a spatial policy as such. Instead it sets an agenda for the then proposed Committee of Spatial Development and identifies four specific areas for EU intervention, namely, frontier developments; co-operation between regions and cities; missing transport, telecommunications and energy links; and the larger cities. Several of these have since been addressed in policy initiatives since Maastricht, as the following chapters show. The main exception is that of urban policy.

The overall agenda for EU spatial policy is based on six themes that emerged from the analysis:

1. Balanced distribution of economic activity, diminishing ties between industry and traditional locations, and dependence of some regions on a narrow industrial base.
2. Demographic change, urban change and implications for urbanisation and inner-city areas.
3. Pressures on transport infrastructure and growing road dependency.
4. Opportunities offered by new developments in ITT.
5. The need to set economic targets within a framework of environmental sustainability.
6. The importance of secure energy supplies to regional competitiveness.

The final point, bringing all of these together, is that a coherent view of the development of EU territory 'requires a new emphasis on planning and cooperation in planning'. It goes on: 'Cooperation between planners, especially at interregional level, is increasingly required', and concludes that the EU 'must have a vision of its future development which makes best use of all its resources' (Commission, 1991: 200).

Transnational studies

Following the completion of the *Europe 2000* report, the second phase of the programme of policy development was a series of transnational studies, commissioned by DG XVI from external consultants, in order to provide a better understanding of EU land use, spatial and regional economic problems. The purpose was to stimulate new ways of thinking about spatial issues which was not inhibited by the mental confines of national boundaries and to encourage bottom-up development of inter-regional links. The set of transnational regions identified for these studies was not proposed as any form of permanent divisions of territory for future policy instruments.

The intention was that each study should cover an area that included parts of at least two member-states, in order to make them truly transnational. One object was to elaborate the material in *Europe 2000* and to develop policy thinking that could lead to a more prescriptive overall spatial planning scheme. Another was educative, to encourage spatial policy-makers to develop the capacity to think across and beyond national boundaries and to identify commonalities which may be susceptible to common policy approaches. The spread of studies included the whole EU, but was designed to ensure that any issues that were specific to certain geographical regions were fully identified so that future policy development for the EU as a whole should avoid the danger of overlooking key issues that only affected, say, Mediterranean or Alpine regions. The initial list of proposed studies was varied somewhat to take into account new political developments, most notably by the addition of a study of the five new *Länder* of Germany, i.e. territory of the former DDR. This is the only exception to the principle that all transnational studies should include parts of more than one member-state. Later studies were also made of areas beyond the territory of the EU.

The transnational study areas were as follows:

- Centre Capitals, containing six national capitals (B, F, D, L, NL, UK).
- Alpine Arc, containing all Alpine and peri-Alpine regions including those of Austria and Switzerland, both of which were candidates for membership at the time the initial list was drawn up (A, CH, D, F, I).
- Continental Diagonal, inland parts of south-west France and north and central Spain (E, F).
- Mediterranean, originally proposed as separate studies of the western Mediterranean and the southern Adriatic regions (E, F, GR, I).
- The Atlantic Arc from the north of Scotland to Portugal (E, F, IRL, P, UK).
- North Sea regions (D, DK, NL, UK).

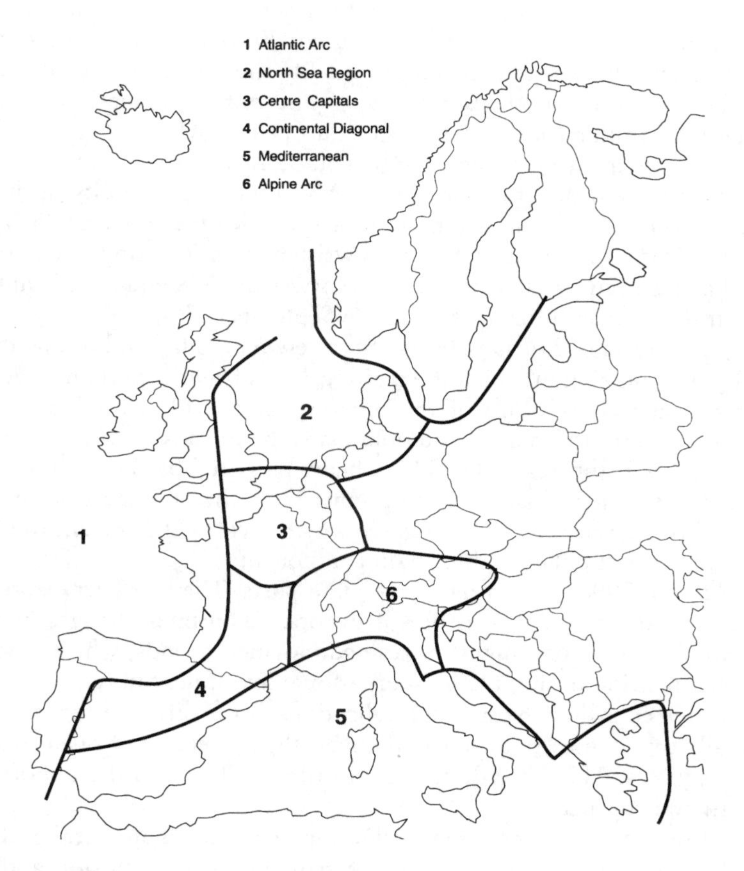

Figure 6.3 *The Transnational Study Areas*

- The ultraperiphery of French overseas *départements*, the Canaries, the Azores and Madeira (E, F, P).

The areas for which these studies were undertaken, apart from the last, are shown in Figure 6.3.

In the initial list, only Austria and Switzerland were included from outside the then EU of 12. This was clearly justified in order to study the Alpine region sensibly. However, the theme of evaluating the respective repercussions of changes in non-EU countries on the territory of the EU and vice versa was extended to include the Nordic countries, three of which were then candidates for EU membership (Norway, Sweden and Finland), the southern and eastern Mediterranean countries, three of which were also candidates for membership (Cyprus, Malta and Turkey), and the countries of the CEE and the western parts of the former USSR.

Europe 2000+

The third stage of this particular sequence of policy studies is the *Europe 2000+* report (Commission, 1994a), which actually came available for distribution in spring 1995. The basic theme of *Europe 2000+* is that growing interdependence between the regions of the EU and increasingly close relations with other parts of Europe requires closer and more systematic co-operation over spatial planning for the territory of Europe as a whole.

It was written at a time when enlargement of the EU was about to occur, but it was strictly the product of the EU of 12 and the responsibility of the Commissioner Bruce Millan whose term of office ended at the end of 1994. The text is based on data for the EU of 12 and is written in anticipation of enlargement to 16, having been drafted before the Norwegian referendum for initial approval at the Leipzig Informal Council in September 1994.

The publication of *Europe 2000+* effectively marks the end of one phase of spatial policy-making in which the main task has been to gain an understanding of the spatial structure of the territory of the EU as a whole and to demonstrate to policy-makers and politicians the purpose and value of spatial policy-making at this scale. The ESDP being prepared by the CSD is intended to lead into the next phase and represents the political extension of *Europe 2000+*. This is discussed in Chapter 12, which reviews the current (mid-1995) state of EU spatial policy and its main components.

The *Europe 2000+* report has three main parts. The first part examines the main trends affecting the territory of Europe. It outlines changes in the distribution of population, migration and employment trends, which is followed by an examination of the role of international investment in regional development. It then considers the spatial implications of the different forms of TENs, and finally two major environmental issues, the protection of open spaces and water resources. These are discussed in order to illustrate the importance of transnational co-operation.

The second part considers major developments in urban, rural and border areas. Urban development problems of growth, social exclusion and spatial segregation, the urban environment, transport and the vulnerability of some small towns are identified as issues requiring specific policy action. The rural focus is on the diversity of trends affecting rural regions and possible ways of responding to the changing patterns of agricultural production and urban–rural relationships. The examination of border areas includes both internal and external EU borders, for both of which co-operation in spatial planning is of especial importance.

The third section is concerned with the spatial planning systems of the member-states, the diversity of planning instruments and policies in force throughout the EU, and their response to the European dimension of spatial planning. It goes on to give some preliminary consideration to the regional effects of public sector financial transfers and taxation. Finally, a summary of the transnational reports referred to above is included in the Annex in order to present a transnational perspective to the different parts of the EU.

Much of the substantive content of *Europe 2000+* is incorporated into the material presented in the context of the sectors of spatial policy in the following chapters (Chapters 7–11); its role as a foundation for further policy development is discussed in Chapter 12, so it is not summarised more fully here.

The Baltic Sea region

A study entitled *Vision and Strategies around the Baltic Sea 2010: Towards a Framework for Spatial Development in the Baltic Sea Region* has been produced on behalf of the Council of Baltic Sea States and adopted by the Council's third Conference of Ministers Responsible for Spatial Planning, held in Tallinn in 1994 (Group of Focal Points, 1994). This study, known by its acronym title as *VASAB 2010*, takes into account all countries with a Baltic Sea coast, plus Belarus and Norway. It thus straddles all the elements of the new European architecture (Chapter 2), including former Soviet countries, Visegrad countries, EFTA and both old and new EU member-states. It also spans extreme disparities of living standards, wealth and environmental quality.

In spite of this, *VASAB 2010* is quite specific in its analysis, its vision for the future and in its proposals for 'common actions' which form the next steps towards an integrated spatial strategy. In many ways, therefore, as a supranational spatial strategy it represents a considerable advance on the efforts of the CoE in the 1970s and 1980s.

The impetus (and the funding) for studies such as *VASAB 2010* comes out of the desire on the part of the rich Baltic Sea countries, the Nordic countries and Germany, to assist in the transformation and economic development of those that were formerly in Comecon, and to guide that transformation so that it does not threaten the environment and fragile ecology of the Baltic Sea region. Recognition of the need for a spatial strategy came out of these considerations. The programme of work commenced in 1992 following an opening ministerial conference in Karlskrona, Sweden.

The study presents an assessment of the spatial problems and potentials of the Baltic Sea region, sets out its vision for the future spatial structure and development of the region, and proposes a strategy and programme of common actions. The organisation of the study is through a group of national focal points, generally an official responsible for international co-ordination of spatial planning policy within each participant country's national spatial planning ministry.

As the report claims, the Baltic Sea region is unique in its multitude of co-operation initiatives, joint programmes and networking activity. This is both a strength and a problem for the VASAB project. VASAB builds on the extensive contact and co-operation, but was initiated by the ministers partly in response to concern that, with so many new links being rapidly developed by many different groups, spatial cohesion has become a priority issue and there is a need to agree common principles and visions for the spatial development of the region.

The first part of the study, analysing the situation being faced, concentrates on four themes: 1) the Pearls (the functions and problems of cities and the urban network); 2) the Strings (transport infrastructure networks, mobility problems, telecommunications, energy supply networks and associated environmental problems); 3) the Patches (areas with specific problems and potentials, including coastal zones, islands, border areas and landscapes of particular cultural or environmental significance); and 4) the System (reviewing the institutional and procedural systems and problems facing the operation of spatial planning in the different countries, and containing practical information concerning the state of development of regional planning authorities and numbers of qualified personnel in the transition of former Comecon countries).

Inclusion of the latter shows clear recognition that, if the project is to be more than just a study, and is to lead to action, then the policy delivery system must be capable of doing the job. In a study area representing such different recent histories and traditions of spatial planning and local and regional government, it is not sufficient simply to develop a spatial planning concept. Availability of expertise and the establishment of an effective system of authorities capable of handling implementation are also essential.

Many of the problems to be addressed under the first three themes are quite predictable, so the emphasis is on identifying issues on which international co-operation and action are required, where local policy-makers could benefit from being able to work within an international context or spatial policy framework, or where external expertise may be helpful. Part of the purpose of the study is to increase international awareness of the problems faced, for example owing to the dramatic differences in living standards and infrastructure across borders, or to the extremes of peripherality in the Arctic and Barents Sea regions. The latter area contains vast oil and gas reserves, which may offer the opportunity to counteract peripheralisation. Another purpose is to educate policy-makers, especially in the transition countries, and particularly in respect of the need to develop improved environmental standards and procedures for environmental protection and impact assessment. The other important educative purpose is in respect of demonstrating the need for spatial planning expertise and suitable forms of local and regional administration, based on principles of transparency, participation and subsidiarity.

The wider vision is that of a region that achieves greater physical and economic integration, develops a spatial structure that meets the highest environmental standards, demonstrates the benefits of spatial policy co-operation and is able to offer practical guidance on integration and harmonisation of spatial aspects of national policies. It is also hoped that the region will as a result be better placed to play a role in spatial policy development for the EU and the whole of Europe, and be recognised as a coherent regional interest within Europe.

The programme of action was agreed in April 1995 to be undertaken or at least started in the first phase in 1995–6, prior to the next meeting of ministers in 1996. Several elements of it involve more detailed studies, including work parallel to the *Compendium of Spatial Planning Systems* project of DG XVI (see Chapter 12), and studies of coastal zone and border region planning and management. The action plan also includes contributing to the EU's ESDP (see

Chapter 12), the work of CEMAT and the CoE (see Chapter 2), and development of a network of spatial research institutes to link to that proposed by the EU.

A direct relationship to EU funding programmes is made in several respects, including use of INTERREG (see Chapter 7) and identification of major transportation corridors that develop the TENs programme eastwards (see Chapter 9).

Perspectives in Europe

The Dutch government's National Physical Planning Agency, the Rijksplanologische Dienst, undertook a study published in 1991 entitled *Perspectives in Europe* (*PIE*) (RPD, 1991). This report analysed, through the elaboration of scenarios, the implications of the different courses which the spatial development of north-west Europe may take. Its territorial extent included Denmark, the former Federal (west) Germany, the Benelux countries, France, the UK and northern Italy.

In the Dutch context, one purpose of the *PIE* project was to explore the spatial context for policy-making at the Dutch national and provincial scales. In this respect, it is a follow-up to the Fourth Report (Ministry of Housing, Physical Planning and Environment, 1988, 1991) which put consideration of the Dutch national space in the European context and indicated that there was a need for spatial planning at the European scale. In spite of the long experience of the Dutch in Benelux and EU co-operation, it was felt that the international dimension required specific study, so the *PIE* project was started in 1989.

Considering the extent to which the port of Rotterdam, Amsterdam Schipol airport and the inland waterway systems of the Rhine and Maas serve European rather than Dutch transport needs, and in doing so sustain a substantial part of the local economy and employment markets, it is not surprising that it was seen as being in the Dutch interest to undertake such a study. In order to explore as widely as possible the European context and range of factors affecting Dutch spatial planning, a scenario approach was adopted. Two scenarios were developed, one based around the growing urban concentration and specialisation of world and European cities, and the other based on trends towards more decentralised urban areas and axes of development. This methodology was later adopted in studies for DG XVI.

The timing of the *PIE* study revealed a second purpose of more significance in the context of promoting acceptance of the idea of a European spatial policy. The schedule was determined by the timing of the Dutch presidency of July–December 1991, the meeting of ministers responsible for spatial planning held in Den Haag in November 1991 and of course the Maastricht Summit of December 1991. The purpose was to demonstrate that it was a practical proposition to develop a spatial strategy at the supranational scale, and that this could form a valuable policy-making tool as a framework for co-ordination of sectoral policies and of national and regional spatial policies of member-states. The ultimate objective therefore was to gain agreement on the inclusion of spatial planning in the revised treaty to give the EU competence.

Benelux

Just as the Benelux Union was in a sense an early prototype for the EU (see Chapter 2), so it may be that Benelux will come to be seen as a prototype for the development of more integrated transnational spatial policy.

The Netherlands has a long history of national spatial policy (Needham *et al.*, 1993; Faludi and van der Valk, 1994) and has been a source of both intellectual and political leadership in support of concepts of supranational and EU spatial policy. Belgium, by contrast, is an example of a country which has had a requirement, in its 1962 national planning legislation, to produce a national spatial plan (Anselin, 1984) which has never been implemented. That requirement no longer applies as a result of regional devolution and the creation of a federal structure. Planning legislation and policy is now totally within the jurisdiction of the Flanders, Wallonia and Brussels regions, so for practical purposes Belgium must be regarded as having not one but three planning systems (Dutt and Costa, 1992).

However, the wider European spatial context is not forgotten in the thinking behind current Belgian planning any more than it is in The Netherlands, nor could it possibly be forgotten by the planning authorities of Luxembourg. Spatial planning, or in its English text, 'town and country planning', has been identified by its Secretary-General as one of the key tasks of the Benelux Union (Hennekam, 1994: 61).

In 1969 the Benelux Committee of Ministers decided that work should start on a spatial plan for the whole Benelux Union, and instituted a special commission for this task (Kruijtbosch, 1987). The first outline structural plan for the Benelux Union was prepared in 1986 and a second has been under preparation during the period 1993–5. This is of interest in the first place as an example of a more detailed spatial policy framework for this vital core region of the EU. It is of wider significance because of the experience gained in developing a strategy which can be implemented through five different planning systems, three of which are based on the relatively new institutional and legal structures of the Belgian regions. The Benelux plan will undoubtedly contribute substantially to the body of experience on which those responsible for formulation of future EU spatial strategies will come to rely.

The objective of the second outline structural plan for the Benelux is to achieve policy co-ordination between the plans and policies of the five planning jurisdictions in respect of four sets of issues:

1. The position of Benelux in the EU.
2. Relationships among Benelux, external frontier zones and neighbouring regions in north-west Europe.
3. The spatial structure of Benelux.
4. Issues requiring collaboration between border regions within Benelux.

On the first two of these, it will state the common position of Benelux; on the other two it will provide a context for other internal and transnational planning projects. It is an attempt to respond to the declaration of the Informal Council of Ministers of Regional Planning in Liège in November 1993 that collaboration between different member-states and regions should play an essential role in the development of an EU spatial development perspective,' and put into practice the call for bottom-up initiatives in order to achieve this.

It aims to provide a context in particular for three crossborder planning initiatives that extend beyond Benelux: the Euroregion of Belgium, Nord-Pas de Calais and Kent; the Dutch–German border areas and the MHAL (Maastricht/Heerlen–Hasselt/Genk–Aachen–Liège) project (see Chapter 8); and the Saar–Lor–Lux project concerning the border regions of France, Germany and Luxembourg. Within Benelux, it will similarly provide the context for the western Scheldt basin and Rhine–Scheldt delta region, the plans for the different regions of Belgium, especially around the Brussels agglomeration, and the Walloon–Luxembourg border area. A selective approach is taken regarding themes on which to concentrate. The main ones are transport corridors and the need for new transport connections (especially in border areas), areas to accommodate urban growth, and rural development (Zonneveld and D'hondt, 1994a).

National spatial plans

It is not unusual for national planning legislation to include a requirement that there should be a national spatial plan. Sometimes, this provision has not been taken very seriously, but during the 1990s there has been increasing interest in the national scale of spatial planning. One reason for this is clearly European integration and recognition of the interconnectedness of spatial development issues across borders. Two examples of national spatial strategies noted in *Europe 2000+*, from Denmark and Germany, are outlined here.

Denmark 2018

One interesting example of this is the study entitled *Denmark Towards the Year 2018* published by the Danish Ministry of the Environment in 1992. The Danish Minister for the Environment is required to present to the Folketing (the Parliament) a report setting out national spatial planning policies of the government. *Denmark Towards the Year 2018*, subtitled *The Spatial Structuring of Denmark in the Future of Europe* (MoE, 1992) is the national planning report presented in 1992 (to be strictly accurate, the English language version is a summary of the Danish original).

Why 2018? The usual explanation is that this year was deliberately chosen in order to provoke the question 'why?', and thus be more memorable among the proliferation of 'Year 2000' studies. Another explanation is that 2018 is 50 years after 1968, by which time it would be safe to assume that anyone who was politically aware or involved in the student movements of that year would no longer be professionally active.

Denmark 2018 is primarily concerned with the spatial structure of Denmark, but the entire strategy and vision is based on Denmark's position in Europe and on the prospect of continuing European integration: 'It is a decisive new angle in national planning to focus strategically on the themes that are also in focus in the European Union. The perspective is based on Denmark's geography and is intersectoral' (MoE, 1994: 12).

In the introduction to the report, the Minister for the Environment, Per Stig Moller, states explicitly that it is a follow-up to *Europe 2000* and to the inclusion in the TEU of 'town and country planning' and 'land use'. He goes on to state that 'the Government of Denmark has an overall goal that Denmark will strive to become the cleanest country in Europe' (MoE, 1992: 5).

More specifically, the spatial development goals are

- Denmark's cities will be reinforced in Europe;
- the Oresund region will be the leading urban region in the Nordic countries;
- Denmark's cities will be beautiful and clean and will function well;
- Denmark's cities will be efficiently linked to the international transport axes in an environmentally sound manner;
- Denmark's landscapes will be varied and the rural areas will flourish;
- Denmark's coasts and cities will keep their distinctive qualities and will be attractive tourist destinations (MoE, 1994: 12).

The first part of the report reviews national planning issues, including urban development, housing need, rural change, the growth of tourism and the development of national and international transport infrastructure. The second part sets out the vision of 2018, prefaced by an imaginary letter written in that year by an admiring foreign visitor. The vision is of a dynamic centre of economic prosperity for northern Europe that can boast environmental quality as one of its international distinctions. The Oresund agglomeration, in particular, is seen as the leading urban region of the Nordic countries. This is the subject of a separate study (see Chapter 8).

The main cities are to be connected to the international transport networks, and links identified for their significance to the changing spatial structure of Europe in accordance with the above goals are emphasised. A high-speed rail link from Berlin and Hamburg via Copenhagen to Oslo and Stockholm, plus improvements on the Esbjerg to Copenhagen route and expansion of Copenhagen's airport, are seen as vital to the strategy. Maritime links and improved harbour access within cities are also proposed in order to realise the vision of an environmentally friendly transport structure.

The third part of *Denmark 2018* sets out the main elements of an action plan. This is a non-binding framework for the county and municipal authorities, and also indicates action by the government and at the EU level. Increasing competition between European cities is recognised, and municipalities are urged to develop their own internationally orientated spatial perspectives and take an active role in joining or forming city networks and co-operation alliances. Concern is expressed that 'Denmark's cities . . . currently tend to be less oriented towards international cooperation and competition than the cities in many other Community countries', so this action is proposed in order to avoid 'letting the cities and regions of Denmark miss the advantages offered by the continuing internationalization' (MoE, 1992: 26). There is an implicit preoccupation with the Blue Banana, expressed in twin concerns to avoid being bypassed by the main transport axes and development forces in the EU, while also avoiding the congestion and pollution characteristic of the central part of the EU.

Denmark 2018 is an excellent example of a country setting out its vision of where it wants to be in the context of the spatial restructuring of Europe that is in any event going to occur. In view of the emphasis on environmental cleanliness, it is also worth reflecting that the Environment Commissioner, Ritt Bjerregaard, was active in Danish politics at the time *Denmark 2018* was prepared.

Framework for the spatial development of Germany

The case of Germany deserves particular attention because it has had the recent experience of integrating two formerly very separate countries following reunification. Secondly, and to some extent as a consequence of this experience, Germany has become a leading advocate of the need for a coherent spatial policy framework at the wider European scale.

The new *Länder*

Faced with the need to develop a policy framework very quickly after reunification, in order to meet the overwhelming political pressure to transform the centrally planned economy through redevelopment and integration into the market economy of the EU, the Federal Ministry for Regional Planning, Building and Urban Development (BMBau in its German abbreviation) prepared the *Spatial Planning Concept for the Development of the new* Länder (*Raumordnerisches Konzept*) in 1991 (BMBau, 1992, English edn). Federal legislation imposes a duty to create equal living conditions throughout Germany, and in its amended version following reunification it emphasises the task of overcoming the east–west divide and creating a spatially balanced settlement structure. In order to avoid delays, a more direct procedure was necessary than was normally the case with the system well established in the western *Länder*.

The spatial planning concept was therefore prepared in order to show the main features of the spatial development strategy for the new *Länder*. It was not comprehensive and was intended to provide a framework for the governments of each of the new *Länder*, who were responsible for its implementation. Particular emphasis was placed on developing several major urban centres and avoiding concentration of development and transport routes on Berlin. This included principles for the distribution of federal and public offices, universities and research institutions among the different urban centres, and policies for rural development and telecommunications.

From the European point of view, a key feature was recognition that this part of Germany would become an important hub for transport and communications to and within central Europe. Attention needed to be given to north–south and east–west growth axes and transport routes. In the case of the north–south axis, the strategy proposed that the Rostock–Magdeburg route was developed to accommodate growth of traffic from Scandinavia in order to avoid overconcentration on Berlin and spread the development pressures. The east–west route was seen very much as part of a wider European strategy to open up the Polish–German border. These spatial concepts were further developed in the traffic route plan for the whole of Germany prepared in 1992.

Guidelines for regional planning

The experience of the spatial planning concept for the new *Länder* encouraged the BMBau to take a positive view of the benefits to be derived from thinking spatially at this scale. The *Raumordnungspolitischer Orientierungsrahmen* (BMBau, 1993, 2nd edn, 1994), abbreviated to *ORA*, was therefore issued in February 1993. The *ORA* is also published in English under the title *Guidelines for Regional Planning*, and in French with summaries in Czech, Danish, Dutch, Polish and Russian (BMBau, 1993). Although it is strictly speaking the case that *ORA* is a national spatial plan, its origin lies in the response to reunification described above, and it was written with the wider European context very much to the forefront. A section of the guidelines is entitled 'General principles for Europe' in which European inter-relations and trans-border co-operation are emphasised, not only within the EU but also with western non-members such as Switzerland and with Poland and the Czech Republic. A sense of spatial positioning is related to Germany's political positioning as a bridge between the EU and CEE countries. The second part of this section sets out principles which Germany believes should form part of an EU spatial policy (see Chapter 12).

Other sections of the guidelines also address the European context whenever appropriate, and also the spatial implications of the removal of the Bundestag and several federal ministries from Bonn to Berlin.

Spatial planning policies in a European context

The importance of a forthcoming presidency as a motivation for undertaking wider spatial planning studies was shown above in the case of the Dutch *PIE* report of 1991. Similarly, in the run-up to their presidency in 1994, the German government sought to promote the idea that the EU should develop its profile in spatial policy.

It did this partly by publicising its own experience and views in the way described above, and by issuing information in a popular format in English as well as German in order to seek general support for the concept of European spatial planning. Germany also promoted ideas for a *European Planning Atlas* at an international conference in Bonn in October 1993, and of an 'Observatory' of networks for spatial research (see Chapters 8 and 12) for the EU and also to assist the countries of the CEE.

One of the brochures produced by the German government for its presidency was entitled *Spatial Planning Policies in a European Context*. Its objective was to explain more clearly the political processes involved in spatial planning for the citizens of the EU, and thereby promote the public acceptability of spatial planning. It is worth quoting the then German Federal Minister for Regional Planning, Building and Urban Development, Dr Irmgard Schwaetzer:

> In establishing the European Union on 1st November 1993 another important step on the road to European integration was taken. Europe is growing closer together as are its towns, cities and regions. The spatial planning of Europe is also changing and being reshaped. The effects of regional planning policy are not felt directly by most citizens. Yet spatial planning concerns us all:
>
> - It includes strengthening the economic and social cohesion of cities and regions;
> - It involves promoting a harmonious development of the community without blurring the identity of cities and regions;
> - And it also aims to reduce the disparities in the stage of development of different regions by improving their attractiveness and strengthening their locational advantages.
>
> (BMBau, 1994a)

This serves as an excellent summary of the position of importance that spatial planning had attained, and illustrates the momentum and political will that is being taken forward into the 1996 IGC.

7

The Cohesion and Structural Funds: the pot of gold at the end of the rainbow

For many people, and many local and regional planning authorities, the EU is seen in the first place as a source of funding. This pot-of-gold mentality is sometimes based on a distorted impression of the scale of EU funding relative to that from national and regional sources. Nevertheless, in respect of spatial policy, the EU budget for the sectors concerned is set to continue growing throughout the decade under the Delors II financial package (see Chapter 5) agreed at the Edinburgh Summit in December 1992, which includes allocation of 15 becus to the Cohesion Fund over the period 1993–9 and doubling the size of the Structural Funds to 176 becus by 1999, when they will account for over one-third of the EU budget.

EU funding is not solely a matter for the Structural Funds, although they are the main focus of this chapter. The Cohesion Fund and the budget for the TENs (see Chapter 9) and some other programmes are also part of the picture.

This chapter therefore includes a look at the overall EU budget before outlining the principles on which the Cohesion and Structural Funds operate. It then goes on to describe their operations in practice. Given the reality of the pot-of-gold attitude, lobbying plays a big part. This is touched on here, but discussed in more depth in Chapters 8 and 13.

Budget

An indication of the scale of the EU budget, and of the provision within it for the Structural Funds and other funding programmes that come within the spatial framework, can be seen from an examination of the budget for 1995 and the proposals for 1996.

The overall budget is divided into broad categories:

1. Common Agricultural Policy.
2. Structural operations:
   - Structural Funds;
   - Cohesion Funds; and
   - EEA financial mechanism.
3. Internal policies.

	Amount (in mecus)	%
Structural Funds	26,579	91.200
of which:		
Community Support Frameworks	(23,253)	(79.800)
Community Initiatives	(3,030)	(10.400)
Transitional Measures and Innovation Schemes	(296)	(1.000)
Measures to combat fraud	(1)	0.003
Cohesion Fund	2,444	8.400
EEA financial mechanism	108	0.400
Total	29,131	100.000

Figure 7.1 *The Structural operations budget* (*Source*: The Preliminary Draft Budget, May 1995).

4. External action.
5. Administration expenditure.
6. Reserves.
7. Compensation.

Structural operations include the Structural Funds and the Cohesion Fund. A total of 29,131 mecus is proposed for 1996, an increase of 10.64% over the 1995 figure. The way in which this sum is proposed to be allocated between funds and objectives is shown in Figures 7.1, 7.2 and 7.3.

As Figure 7.1 shows, the bulk of the Structural Funds (87.5%) is allocated to the Community Support Frameworks agreed for the different regions designated to benefit. Part of this budget is allocated within single programming documents. These are explained and examples described below. Community Initiatives account for 11.4% of the Structural Funds budget and are also explained below. The final 1.1%, for Transitional Measures and Innovation Schemes, is sometimes referred to as the Article 10 money because the legal basis is Article 10 of the ERDF regulation. This part of the budget funds some of the initiatives central to the development of an integrated spatial policy for the EU. It has funded the *Europe 2000* and *Europe 2000+* studies (see Chapters 6 and 12) and many of the RECITE projects and networking activities described in Chapter 8.

The allocation of the 23,253 mecus for Community Support Frameworks is shown by objective in Table 7.2 and by fund in Table 7.3. The Financial Instrument for Fisheries Guidance (FIFG), for restructuring areas suffering from the decline of the fisheries industry, is strictly speaking an instrument not a fund. Therefore it is not one of the three Structural Funds although it is

	mecus	%
Objective 1	15,424	52.9
Objective 2	2,634	9.0
Objective 3	2,216	7.6
Objective 4	391	1.3
Objective 5(a)	1,115	3.8
Objective 5(b)	1,332	4.6
Objective 6	141	0.5
Total	23,253	79.7

Figure 7.2 *Budget for Structural Funds by objective* (*Source*: The Preliminary Draft Budget, May 1995).

	mecus	%
ERDF	11,884	40.8
ESF	7,146	24.5
EAGGF guidance	3,772	12.9
FIFG	450	1.5
Total		

Figure 7.3 *Budget for Structural Funds by fund* (*Source*: The Preliminary Draft Budget, May 1995).

included in these figures because in the 1994–9 programming period FIFG expenditure is co-ordinated in the same way as the Structural Funds through the Community Support Frameworks.

Economic and social cohesion

The pursuit of economic and social cohesion underlies many spatial policy initiatives: indeed, Article 2 of the TEU makes this duty explicit. Title XIV of the TEU, 'Economic and social cohesion', sets out the overall goals of the Structural Funds and other existing funds: 'the Community shall aim at reducing disparities between the levels of development of the various regions and the backwardness of the least favoured regions, including rural areas' (Article 130a). Also in Title XIV, Article 130d committed the EU to the creation of a Cohesion Fund by the end of 1993.

Cohesion Fund

Formal inauguration of the Cohesion Fund was not until 25 May 1994 because the delays in ratification to the Maastricht Treaty also delayed formal adoption of the regulation establishing the fund. Operations did begin in 1993, however, on the basis of an interim financial instrument adopted in March 1993 and the agreement on the budget and allocation to the four beneficiary countries reached at the Edinburgh Summit in December 1992.

The Cohesion Fund provides funding which contributes to economic development through transport and environmental improvement projects. These are not its central purpose, however, only the means to an end. The fundamental reason why it was written into the treaty as a commitment was to enable all member-states to join the final phase of economic and monetary union as quickly as possible by helping those with the greatest number of handicaps to overcome them. The Protocol on Economic and Social Cohesion annexed to the Maastricht Treaty includes the agreement that the Cohesion Fund will provide financial contributions to projects in the fields of environment and TENs to member-states with a per capita GNP of less than 90% of the EU average which have adopted programmes to achieve the conditions of economic convergence agreed elsewhere in the treaty under Article 104c.

Four member-states meet these conditions – Greece, Ireland, Portugal and Spain – sometimes referred to colloquially as the 'poor four'. All are on the southern or western periphery of the EU. Spatially, the Cohesion Fund will transfer quite substantial resources to these parts of the EU. In terms of GNP, these four form a distinct group. In 1993, their GNP per capita as a percentage of the average for the anticipated EU of 16 was, respectively, Greece 49%, Portugal 60%, Spain 76% and Ireland 78%. The UK was the next poorest of the then 12 member-states at 99%, while Finland at 86% and Sweden at 98% were also below the average (Commission, 1994b).

The Cohesion Fund does not operate in isolation from other funding programmes although it does have its own distinct features. Since its purpose is to reduce disparities between member-states, it is only indirectly contributing to regional policy within the four beneficiaries. Unlike the Structural Funds, it operates on a project-by-project basis, projects being agreed between the Commission and the member-state governments. Funds are available for environment and transport infrastructure projects only. In effect, the Cohesion Fund is the principal financial instrument associated with the programme of TENs (see Chapter 9).

The regulation establishing the fund lays down indicative allocations of funding among the four countries that are reminiscent of ERDF quotas:

Greece	16–20%
Ireland	7–10%
Portugal	16–20%
Spain	52–58%

Structural Funds

The three Structural Funds are the ERDF, the ESF and the guidance section of the EAGGF. Their evolution and co-ordination was outlined in Chapter 5. They continue to play an enormous role in the strategy of economic and social cohesion. Unlike the Cohesion Fund, the spatial designation of areas of benefit from the Structural Funds is based in subnational divisions of territory rather than inclusion of whole countries.

NUTS units

In order to define the areas of benefit of the Structural Funds, and indeed of any spatial targeted policy instrument, a system of territorial definition that can work throughout the EU is necessary. The basis of any such definition must of course be the maps of local and regional authority boundaries of each member-state. However, the systems of subnational government and the terminology used for each level of the hierarchy in each country are so varied that some common definitions became necessary.

The system used was devised by the Statistical Office of the European Communities, in the first place in order to provide a common classification of levels of subnational territorial divisions employed for the collection of statistical data. This is known as the NUTS system, or Nomenclature of Territorial Units for Statistics, the acronym being based on the French version of this title.

The whole of the EU is classified into three NUTS levels. Each member-state is divided into a whole number of level-1 units, each of which in turn is divided into a whole number of level-2 units, themselves divided into a whole number of level-3 units.

NUTS units are based primarily on institutional subdivisions currently in force in member-states. As far as possible the NUTS classification uses existing general purpose units, rather than those serving specific and possibly temporary purposes. Although functional or homogeneous regions may be more truly comparable for purposes of statistical analysis, this approach was rejected for practical reasons of data availability and the implementation of spatially targeted policies, such as the Structural Funds.

NUTS-1 units are regions or federal states where these are large (e.g. Belgium, Germany), autonomous regions (e.g. Spain), groups of smaller regions (e.g. Italy), and they may not be jurisdictions (e.g. UK standard regions). In the cases of Denmark, Ireland and Luxembourg, the whole of the country counts as one region at NUTS-1 level, and also at NUTS-2 level. NUTS-2 units are provinces (e.g. Belgium, The Netherlands), smaller regions (e.g. France, Italy) or groups of countries (e.g. the UK). NUTS-3 units include French *départements*, Irish planning regions, Spanish provinces and UK counties. At this level, Luxembourg is the only undivided case.

The EU of 12 member-states had 71 NUTS-1 units, 183 NUTS-2 regions and 1,044 NUTS-3 units. Classification of the subnational divisions of the three new member-states was not finalised at the time of accession (Figure 7.4).

Correspondence Between NUTS Levels And National Administrative Divisions In The Community						
	NUTS 1		NUTS 2		NUTS 3	
B	Régions	3	Provinces	9	Arrondissements	43
DK		1		1	Amter	15
D	Länder	16	Regierungsbezirke	40	Kreise	543
GR	Groups of development regions	4	Development regions	13	Nomoi	51
E	Agrupacion de comunidades autonomas	7	Communidades autonomas + Melilla Y Ceuta	17 1	Provincias	50 2
F	ZEAT + DOM	8 1	Régions	22 4	Départements	96 4
IRL		1		1	Planning regions	9
I	Gruppi di regioni	11	Regioni	20	Provincie	95
L		1		1		1
NL	Landsdelen	4	Provincies	12	COROP-Regio's	40
P	Continente + Regioes autonomas	1 2	Commissaoes de coordenaçao regional Regioes autonomas	5 2	Grupos de Cancelhos	30
UK	Standard regions	1	Group of counties	35	Counties / Local authorities areas	65
EUR12		***71***		***183***		***1044***

Figure 7.4 *The NUTS levels and national administrative divisions* (*Source*: Commission, 1994b: 173).

The 1993 revisions to the Structural Funds

The recommendations put forward in the Delors II package (Commission, 1992a) for the Edinburgh Summit were based on the four key principles underlying the 1988 reform of the Structural Funds: concentration, programming, partnership and additionality. The system of co-ordination of the Structural Funds around a set of overall objectives will continue at least until 1999, but revised regulations adopted for the programming period 1994–9 introduced a number of changes designed to improve their effectiveness. The lists of eligible regions was also revised.

The main changes were as follows:

- Integration of action in the fisheries sector and in areas dependent on fisheries.
- Creation of a new Objective 4 designed to facilitate the adaption of workers to industrial change and changes in systems of production.
- Widening of the range of actions in Objective-1 regions to include education and health measures.
- Simplification of decision-making.
- Involvement of the social partners in decision-making.
- Strengthening of checks to verify additionality.
- More precise formulation of quantitative objectives and greater emphasis on appraisal and evaluation.
- Indicative allocations to member-states decided by the Commission.

The five priority objectives agreed in the regulation for the Structural Funds for the period 1994–9 are set out in Figure 7.5, together with Objective 6 which

Objective 1 Economic adjustment of regions whose development is lagging behind

Objective 2 Economic conversion of declining industrial areas

Objective 3 Combating long-term unemployment and facilitating the integration into working life of young people and of persons exposed to exclusion from the labour market

Objective 4 Facilitating the adaptation of workers to industrial changes and to changes in production systems

Objective 5(a) Adjustment of the processing and marketing structures for agricultural and fisheries products

Objective 5(b) Economic diversification of rural areas

Objective 6 Economic adjustment of regions with outstandingly low population density

Figure 7.5 *The Structural Fund Objectives*

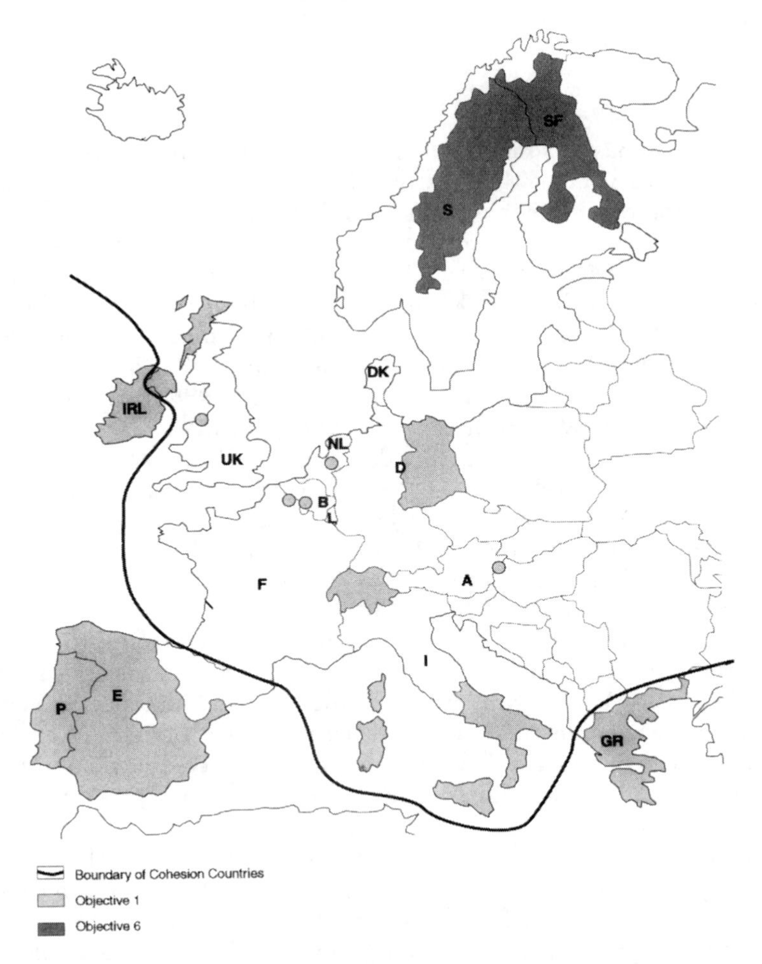

Figure 7.6 *Areas designated under Objectives 1, 6 and the Cohesion Fund*

was added through the negotiations for the 1995 enlargement. Objective 6 is in principle an Arctic version of Objective 1, for regions defined at NUTS-2 level in Finland and Sweden (an allocation was also made for Norway) in the north with population densities below eight persons per square kilometre. The legal basis is a Protocol to the Treaty of Accession, but Objective 6 will be fully integrated with the Structural Fund regulations at their next revision in 1999.

Objectives 1, 2, 5(b) and 6 have explicit territorial definitions, discussed below. Objectives 3 and 4 are primarily financed by the ESF and apply throughout the EU, so they are not of particular spatial policy significance although they form a vital component of the development strategies set out in the Community Support Frameworks. Elsewhere local and regional authorities also incorporate them into their economic development programmes. Objective 5(a) is a horizontal measure funded by the guidance section of the EAGGF.

Like Objectives 3 and 4, it applies throughout the EU and is not discussed further here.

Objective 1

Greatest interest has surrounded eligibility under Objective 1 (Figure 7.6) – not surprising in the light of the growing share of the funds allocated to these areas. This follows from the Commission's view that resources need to be concentrated in the worst affected areas in order to be most effective in reducing disparities. Between 1989 and 1993, the proportion of funds going to these regions rose from 62% to 65%, and by 1999 the proportion including the Cohesion Fund will reach 73%. The four poorest member-states received 42% in 1988, 50% in 1992 and should reach 54% in 1999. The share for other Objective-1 regions will rise from 19% to 23% by 1999. Expenditure on the other objectives will fall in relative terms although it will rise in actual volume.

The basis criterion for Objective-1 designation is that it should be for areas defined at NUTS-2 level with GDP per head of less than 75% of the EU average. However, some flexibility has been applied by Council to reflect political considerations. Thus, in 1988, Northern Ireland and Corsica were included although they did not meet the 75% GDP criterion. For the 1994–9 period, these two plus six other areas were added that did not meet this criterion: Abruzzi and Molise in Italy, Hainault in Belgium, part of Nord-Pas de Calais in France and Merseyside and the Highlands and Islands Enterprise Board area in the UK. New additions to the list which did meet the criterion were the five new German *Länder* and the eastern part of Berlin, Cantabria in Spain and Flevoland in The Netherlands. All of Greece, Ireland and Portugal are included, while only Denmark and Luxembourg of the pre-1995 EU of 12 now have no Objective-1 areas.

No region was taken off the Objective-1 list in 1994 although Abruzzi, the most prosperous, will lose its status on 1 January 1997. Governments naturally would resist removal from Objective 1 in view of the scale of resources concentrated on it and their significance to national budgets. Several national and regional authorities will therefore watch carefully to see whether their GDP per head is reaching the upper limits of eligibility as the next review comes due. Ireland arguably benefits from the fact that the entire country is one region at NUTS-2 level. If this were not so, it is possible that the most prosperous region, the Dublin region, might be proposed as Objective 2 instead.

The areas added in 1994 extended the type of area included in Objective 1 in two ways. The former Objective-2 areas added in Belgium, France and the UK are not areas of lagging development in the traditional sense of never having been highly industrialised, but are areas where the decline in relative prosperity based on industrial activity has been especially acute.

The second new situation is that of the new German *Länder*. Of the extra population added to Objective-1 areas in the review, 75% is accounted for by the 16.4 million inhabitants of the new *Länder*. Now over 20% of the population of Germany is in Objective 1, whereas previously Germany was not included. As noted earlier, the particular form the regional problem takes there (transformation from a centrally planned economy to a market-orientated mixed economy) is without historical precedent. Implementation will offer valuable experience with which to prepare for future enlargements.

New member-states

From the three new member-states, only Burgenland in Austria, alongside the Hungarian border, qualifies for Objective-1 status. During the enlargement negotiations, the three Nordic candidates argued that conditions in the remote Arctic regions warranted inclusion in Objective 1, but this was unacceptable as it would have meant deviating too far from the criterion. The compromise was the new Objective 6, defined on the basis of population density but implicitly acknowledging that remoteness and the climate justified assistance if they are to avoid lagging behind economically. In Sweden, the three northern counties are eligible, and in Finland the area includes Lapland and border areas with Russia.

Objective 2

The proportion of the population included in Objective-2 areas of the EU of 12 remained much as before at 16.8%. In general, the number of individual designated areas has increased while their average size has decreased. The list agreed in 1994 is effective for only three years, 1994–6, and eligibility is to be reviewed in 1996 for the following three years, 1997–9.

Objective-2 areas are defined on the basis of NUTS-3 regions, or parts of them. Individual communes, *Gemeinden*, parishes or travel-to-work areas (TTWA) may be added or deleted from NUTS-3 units in order to arrive at the final list of eligible areas. The pattern of designation is therefore much more complex. The full list for 1994–6 for the EU of 12 is 52 pages long and is published as a Fact-sheet Annex. This gives an indication of the scale of Objective 2 (Commission, 1994c).

Objective-2 areas are identified from NUTS-3 areas, or parts of them, which meet the following basic criteria:

- Unemployment rate above the EU average.
- Proportion of industrial employment not less than the EU average.
- Declining trend in industrial employment.

Some flexibility is available to include areas such as those

- adjacent to Objective-1 areas;
- urban communities with unemployment rates at least 50% above the EU average and a sharp downward trend in industrial jobs;
- suffering or liable to suffer substantial job losses in industries decisively important to economic growth;
- affected by restructuring of the fishing industry; and
- especially urban, facing serious problems in the rehabilitation of derelict industrial sites.

Apart from Greece, Ireland and Portugal, which are entirely within Objective 1, all member-states proposed areas for inclusion in Objective 2 in late 1993. As the aggregate of all proposals amounted to 22.5% of the EU-12 population, well exceeding the indicative figure of 15% specified in the Structural Funds regulation, 'close consultation', a euphemism for hard negotiations, took place between the Commission and national governments before the list was agreed early in 1994. The figure of 16.8% represents the compromise reached (*OJ* L81 24.3.94). The extent of Objective 2 in Austria and Sweden is quite small, but rather more of Finland is designated.

Objective 5(b)

Objective 5(b) regions have been defined for the full six-year period on the argument that structural change in rural areas takes place relatively slowly, and because rural areas face new challenges from CAP reform as well as from the effects of the GATT Uruguay round agreement. The total population of the EU of 12 in Objective 5(b) areas has increased from 5% to 8%, the largest rises occurring in Denmark and Luxembourg (*OJ* L96 14.4.94).

Programming and planning expenditure of funds

Eligible actions to contribute to the achievement of these objectives fall into three categories: those taken under Community Support Frameworks (CSFs), those based on Single Programming Documents (SPDs) agreed between the Commission and member-states on the basis of their own national or regional development plans, and Community Initiatives. In terms of the overall EU budget (see above), allocations made in SPDs are accounted under the heading CSF.

Community Initiatives, along with Innovative Measures under Article 10 of the Structural Funds regulation, use a proportion of the funds that the Commission is able to employ on its own initiative. Expenditure under the latter two categories is therefore most likely to reflect truly European spatial policy perspectives, at least in the judgement of the Commission.

SPDs are a new approach to the administration of the Structural Funds, introduced for the 1994–9 programming period. In a sense, they are the successor to the NPCI concept, whereas the CSFs are the successors to Community Programmes. However, for budgetary purposes, SPDs are included in the same category as CSFs. The bulk of Structural Fund expenditure takes place within CSFs.

The Structural Funds do not meet the total costs of the projects they support. All structural funding is on the basis of partnerships, with funding required also from national, regional and local authorities in the member-states concerned. The EU funds can support up to 85% of costs in the remotest regions and certain Greek islands, up to 80% in Objective-1 regions in Greece, Ireland, Portugal and Spain, up to 75% in other Objective-1 regions and up to 50% cofinancing of actions under Objectives 2, 3, 4, 5(b) and 6.

CSFs

The CSF is a key document, being an agreement between the Commission and the member-state containing specific financial commitments from the EU. It sets out how the Structural Funds, plus EIB and ECSC finance if appropriate, will be deployed to assist member-states achieve their regional planning objectives.

CSFs are, however, only one stage of an elaborate regional economic planning and decision-making process for the Structural Funds, in which local and regional authorities, national governments and the Commission all participate.

The first stage involves preparation within the member-state of a regional development or conversion plan, containing priorities identified at national, regional and local levels, for each Objective 1, 2 or 5(b) area. These are submitted to the Commission, and detailed negotiations take place leading to the preparation of a CSF for each region or for the member-state as a whole.

The bulk of the allocations in a CSF are likely to take the form of Operational Programmes, preparation and approval of which is the next stage in the process. CSFs can also provide for Global Grants to designated authorities, project finance and part-financing of national regional-aid schemes. Operation Programmes are essentially schemes for implementation of the CSF. These also have to be agreed with the Commission. Regional and local authorities are then responsible for proceeding with implementing the programme.

This procedure is not intended to operate in a totally centralised manner. It is designed to allow for a greater degree of local autonomy than was the case under earlier ERDF procedures which required project-by-project approval by the Commission. Local and regional authorities are able to decide which individual projects to fund in order to meet the agreed overall objectives of the Operational Programme. This accords with the Commission's view of subsidiarity, whereby the greater the degree of decentralisation, the greater the likelihood that the most appropriate action will be taken to suit local circumstances. It also means that the authorities most closely concerned should take responsibility for local decisions.

Additionality and transparency are two other important key words. It has consistently been the Commission's ambition to enforce the principle of additionality, which requires that EU assistance should be additional to, rather than a substitute for, national and regional funding. Just as consistently, member-states sometimes find it convenient to be less than strict about the additionality principle. Hence, there is also an emphasis on the need to improve the transparency of national accounting procedures so that the sources of investment finance, whether national, regional or EU, can be clearly identified.

SPDs

The SPD is a rather simpler single-stage procedure, combining the functions of the CSF and Operational Programme, so enabling the agreed programme of action to get under way as quickly as possible. They are drawn up on the basis of proposals from the member-states concerned, and agreed by the Commission after negotiation. Although approval depends on meeting EU criteria, SPDs may give greater emphasis to national rather than EU priorities.

The SPD procedure has been adopted in some of the newly created Objective-1 areas, particularly those that are relatively small in area, those whose problems are capable of treatment in the national and regional context, or where the degree of integration necessary with other EU programmes is limited. Examples where the SPD procedure has been used include Merseyside and the Highlands and Islands (UK) and Flevoland (NL). The SPD procedure is also used for Objective 6, the example of Finland's SPD being outlined below, and it can also be adopted for Objective 2 and 5(b) areas, one example of the latter being the northern uplands of Durham and Northumberland (UK).

Examples of Objective-1 programmes

In order to give a fuller picture of the range of policies and interventions funded through the CSFs and SPDs in different member-states, some examples are outlined here. The examples selected include two CSFs from countries that are wholly Objective 1, Ireland and Portugal; two from other Mediterranean countries, Italy and Spain; plus the distinct case of the new *Länder* of Germany. The two SPDs are Merseyside and Flevoland.

The CSF for Portugal

Since Portugal joined the EU in 1986, it has been a major beneficiary of the Structural Funds. At one time it was hardly possible to pass a road junction without seeing a notice indicating that improvements had been, or were to be, undertaken with the support of the ERDF.

Considerable economic development has taken place, bringing Portugal closer to the economic level of other member-states. The richest mainland region, Lisboa e vale do Tejo, has a per capita GDP of 76.6%, while the figure for the poorest, Alentejo, is 33.9%. Since economic convergence still has some way to go, the whole of the country is included, as are the autonomous regions of the Azores and Madeira.

The CSF lists a number of problems and obstacles to economic development which, it is argued, require public sector intervention on a scale beyond the resources of the national authorities. These include the small size and lack of specialisation of the national markets, low levels of education and vocational skills, weak industrial fabric, inadequate infrastructure, remote location within Europe, conditions for agriculture, regional disparities and underdevelopment of the urban network.

Four overall priorities for assistance have been set:

1. Developing human resources and promoting employment through education and training programmes, allocated 15.5% of the budget.
2. Improving economic competitiveness through the development of infrastructure, especially transport, energy and telecommunications networks, and modernisation of key economic sectors including agriculture, fisheries, tourism, retailing and manufacturing industry, allocated 59% of the budget.
3. Improving the quality of life and social cohesion through environmental action, urban renewal, and improvements to health and social provision, allocated 7% of the budget.

Notice for ERDF-funded industrial development project in Portugal. (photo: the author)

4. Strengthening the regional economic base through programmes of local and rural development, support for local authorities, and specific programmes of regional aid for each region, allocated 18.5% of the budget.

A total of 16 Operational Programmes have been agreed on the basis of the CSF (a simplification of the situation during the 1989–93 funding period when there were 60 of them). Six are sectoral, for sectors such as the knowledge base, vocational training, infrastructure, industrial development, environment and urban renewal, and social integration. There are seven regional Operational Programmes, one for each region, and the others are for a national programme for regional assistance, for modernisation of the official statistical system and a global grant to support local-authority development projects.

Since the Lisboa e vale do Tejo is already near the upper limit of the range of per capita GDP accepted for designation as Objective-1 regions, any improvement in the position achieved by the 1994–9 programme may call into question this designation. Since the two immediately above on the listing in the fifth periodic report (Commission, 1994b) are Merseyside at 76.7% and Highlands and Islands at 76.9%, inclusion of the latter may have helped confirm the Objective-1 status of Lisbon, and their situations will be linked in any future review.

Ireland

Ireland has some problems in common with Portugal, such as a high dependence on agriculture and the need for communication infrastructure to overcome the effect of its peripheral location. Several differences are noted, such as the relatively high educational level and extent of non-Irish industrial investment. Unemployment increased during the 1989–94 period, as new job opportunities were not sufficient to overcome the effects of demographic growth and reduced emigration rates.

Four priorities for the CSF are

1. development of the productive sector, including agriculture, forestry, fishing, tourism, manufacturing and food-processing industries, allocated 38.9% of the budget;
2. economic and communication infrastructure, energy and environment, allocated 19.8%;
3. human resources, education and social integration, allocated 37.6%; and
4. assistance to local enterprises in urban and rural areas, allocated 3.6%.

Within the second of these, transport priorities include the development of the three international airports at Cork, Dublin and Shannon, and of the ports of Cork, Dublin, Rosslare and Waterford, plus an appraisal of the rail network. Telecommunication investment includes fibre-optic cabling and automation. The environmental programme includes water recycling, waste disposal, coastal protection and monitoring. The major social infrastructure proposal is for a new teaching hospital at Tallacht, west of Dublin.

Notice for ERDF-funded road project in Ireland. (photo: the author)

The detailed development strategy and justification for the CSF is set out in Ireland's *National Development Plan 1994–1999* (Government of Ireland, 1993), which also reviews the outcome of the 1989–93 CSF and provides a framework for implementation through the Operational Programmes and the Community Initiatives (see below) in which Ireland is participating.

Co-operation with Northern Ireland (also an Objective-1 area) is an important element which has received substantial EU backing since the peace process of 1993–4 gained momentum.

Ireland is classified as one region of the EU at NUTS-2 level, and consequently the whole of the country is treated as one region for the purposes of the Structural Funds. Without a change in the basis for designation of Objective-1 areas, it would therefore be more difficult than in the case of Portugal to separate the most prosperous region, the Dublin area, from the rest of the country in any future review of Objective-1 designation.

Spain

Spain, like Ireland and Portugal, is a Cohesion Fund country, but unlike them not all its area has Objective-1 status. The areas that do not are the Madrid area, regions along the French border including Cataluna, Aragon and the Basque country, plus the Balearic Islands. The remainder, amounting to 77% of the area of Spain including the Canary Islands and two small pieces of Spanish territory on the African mainland, Ceuta and Melilla, is Objective 1.

Weaknesses diagnosed include problems with the levels of new technology and innovation in productive sectors of the economy, lack of vocational education, remoteness within Europe and problems of communication within Spain, environmental problems and non-sustainable economic processes, and inadequate social provision.

Four priorities have been set. To

1. improve the production system (34.5%);
2. develop human resources through education, training and health services (33.4%);
3. reduce isolation of remote regions through development of the road network, improvements to the rail system including conversion of certain routes to the European standard gauge, port and seaport development and telecommunications (24.8%); and
4. develop energy networks and water management (7.3%).

The 1994–9 programme gives more emphasis than earlier programmes to production systems and human resources, and rather less to reducing isolation because of improvements that have been achieved. Additionally, Spain stands to benefit from the expenditure of Cohesion Funds on TENs. Environmental objectives are integrated into all economic programmes for the first time in Spain through a *national strategic plan for the environment*, to which the CSF is related. There are 18 Operational Programmes, 14 of which are explicitly regional. Overall, greater regionalisation of investment is sought. Investment of 33.8% of the total CSF is directly allocated to regional administrations, and a large proportion of the nationally controlled CSF budget is also allocated to programmes which are implemented within the framework of a regional strategy. In this way, a total of 90% of the CSF is to be expended on a regionalised basis.

Italy

In EU terms, Italy is a country of great contrasts. In the prosperous north, GDP per capita exceeds that of France and Germany, whereas the south or Mezzogiorno has a per capita GDP less than that of Spain. The CSF has identified a number of factors behind this situation, which it seeks to address. These include, in the rather euphemistic language of the Commission, an inefficient production system, shortage of business services, a very high level of public and private consumption compared with production linked to an alarming public-spending deficit, public spending used to protect incomes rather than capital investment, inadequate public services, lack of control over public spending, and inadequate basic infrastructure. Its remote geographical position is also noted.

Eight priorities have been set for 1994–9:

1. Communications: road and rail improvements to be linked into TENs, and telecommunications (14.5%).
2. Industry, SMEs, craft businesses and business services, industrial estate development (25%).
3. Tourist development and the natural and cultural environment (5.8%).
4. Diversification and development of agricultural resources and rural development (15.7%).
5. Fisheries (1.7%).
6. Infrastructure to support economic activity, water resources, energy networks, renewable energy, water and waste management, research and development (22.2%).
7. Development of human resources through education, programmes for the long-term unemployed, training for public administrators (14.5%).
8. Technical assistance, publicity and monitoring (0.6%).

The number of Operational Programmes is to be kept as low as possible to simplify implementation, but the list is likely to include Operational Programmes for each main sector, plus one for each of the eight regions involved. In the case of Abruzzi, east of Rome, this only covers the period 1994–6, after which it reverts to Objective-2 status.

Some convergence and increased selling of goods and services from enterprises supported by the 1989–93 CSF in the north have taken place. Among educational investment were new faculties of engineering and architecture in the University of Reggio Calabria, a university which is noteworthy in the context of planning education as it offers one of the first two spatial planning degree programmes to be offered in Italy.

As part of the REGEN Community Initiative (see below and Chapter 8), the Italian and Greek electricity networks are being connected from Porto Badisco in Apulia.

The new German *Länder*

In contrast to Italy, which has benefited from EU regional funding in its various guises for as long as it has existed, the new *Länder* have only been fully integrated into the Structural Fund objectives for the 1994–9 period. The German Objective-1 area is the entire territory of the former GDR, i.e. the five new *Länder* plus the eastern part of Berlin. The problems posed are unprecedented, as is the scale of the disparity with the rest of Germany. NUTS-2 regions in this area were not agreed when the fifth periodic report was compiled, but the five new *Länder* all fall within the bottom seven for the whole EU of 12, ranked by GDP per capita in 1991 (Commission, 1994b: Table A.25). Many projects were supported by the three funds during the period 1991–3. ERDF funding was directed at production-related infrastructure, mainly in the form of site preparation, and in support for the establishment of new SMEs. The ESF was directed at training and alleviation of the effects of the rapid rise in redundancies.

Nevertheless, a long list of weaknesses has been identified by the CSF, several of which would apply with greater intensity to any of the Visegrad countries. These include

- lack of private investment and of investment funds;
- lack of a high-technology base;
- problems with creation of SMEs;
- poor-quality telecommunications and basic transport infrastructure;
- massive job shortage as overemployment is eliminated;
- agriculture not market orientated or environmentally acceptable;
- lack of alternative economic base in the extensive rural areas; and
- widespread pollution and contamination in areas with economic or tourist potential.

Six priorities have been established:

1. Productive investment (18.2%).
2. Financial and service support for SMEs (17.4%).
3. Research, technological development and innovation (4.6%).
4. Protection and improvement of the environment (8.3%).
5. Measures to combat unemployment, and to promote human resource development and vocational training (27.3%).
6. Measures to promote agriculture, development of rural areas and fisheries (24.2%).

Flevoland and Merseyside

The SPDs for these two areas are set out in a similar way to the CSFs, with an analysis of the economic situation and a set of priorities to which a proportion of the budget as allocated. Although they are for very different types of place, they are both for places whose problems are well researched and understood, and where the general educational levels and institutional structures are capable of supporting a sophisticated planning and implementation process.

This is one reason why the SPD approach rather than the two-stage CSF and Operational Programme process was preferred in these cases.

Flevoland consists of reclaimed polders, part of which is high-quality agricultural land which has been farmed very intensively and efficiently, but which now faces problems owing to the fall in agricultural prices and reforms of the CAP. Much of the urban development that has taken place in Flevoland, especially in Almere and Lelystad, has accommodated overspill population from the Randstad cities. Many residents work in the Randstad, giving Flevoland a net commuting outflow corresponding to 20% of the region's workforce. The added value they produce is included in the GDP of the adjacent areas. The aim therefore has to be to create additional jobs and achieve a better balanced employment structure to avoid overdependence on the agricultural sector.

A total of eight priorities were set out, within which four drivers for change were identified:

1. The regional business sector, in particular SMEs.
2. Tourism.
3. Agriculture, the agribusiness sector and rural redevelopment.
4. The fisheries sector.

A set of specific programmes of activity are elaborated in the SPD in order to support each of these drivers.

Merseyside is a commercial and industrial city region whose former prosperity was closely connected with its role as a major Atlantic port. To some extent its location was itself a handicap as intra-EU trade became a more important feature of the UK economy since membership in 1973, favouring east-coast ports. However, Merseyside suffered many other political and economic problems in the 1980s and during that decade its per capita GDP fell from 86% to the 1991 figure of 76.7%. It had Objective-2 status in 1989–93 and in the 1993 review its position was on the margin of Objective 1 (see Figure 7.8). Apart from comparisons with other regions, its figure of 25% male unemployment was a critical factor in tipping the scales in favour of Objective 1 (Boland, 1995).

Five priorities are set out in the SPD:

1. Actions to promote inward investment and business development, including site preparation and improvements in access and communications.
2. Actions to promote indigenous enterprises and local firms, focusing on premises for SMEs, and measures to support their financial, telecommunications and energy needs.
3. Development of advanced technology and knowledge-based industries, product development and technology transfer from the universities.
4. Action to develop the cultural, media and leisure industries.
5. Measures to assist the people of Merseyside through community development in inner-city areas, basic literacy, numeracy and job-seeking skills, training in industrial and technical skills and work-experience projects.

An unusual feature of the Merseyside SPD is the large proportion of ESF expenditure, reflecting the severity of the local unemployment situation. Under the original proposal, the ESF accounted for 47% of total expenditure, and in the agreed SPD it is 41%. This is still the largest proportion allocated to the ESF of any Objective-1 region (Boland, 1995).

Objective 6 in Finland

The Arctic regions of Finland and Sweden represent a new climatic and geographical situation to which established EU policies have to be adapted. These regions are characterised by a cold harsh climate, a very short growing season and very sparse population density. Objective-6 designation is available to areas of under eight persons per square kilometre and applies only in Sweden and Finland. The example of the Objective-6 regions of Finland, where the average population density is in fact four persons per square kilometre, is taken to illustrate this element of EU spatial policy.

The designated area forms the north-eastern border of the EU. It has a border of over 1,000 km with Russia to the east and a border with Norway to the north. Had Norway joined the EU, this part of Norway would also have been an Objective-6 area. A total of 60% of Finland's surface area is covered, although the population of 840,000 amounts to only 16.6% of the national total. Five thousand of these are Sami or Lapps, whose traditional way of life is an important part of the cultural heritage and is legally protected, as it is in Norway and Sweden as well. No towns in the area are large, the only three of any size being Joensuu (50,000), Kajaani (37,000) and Rovaniemi (34,000). In the southern part there is a scattered settlement pattern, while in the north it is more concentrated along rivers. There has been a steady loss of population during the last 30 years, accentuated by the loss of young people and a low birth rate.

The standard of transport infrastructure in the form of road, rail and air links is good, and the level of telecommunications is well advanced although the availability of new services falls below the high standards of the south of Finland. The environment presents more of a problem. The majority of the area is pristine wilderness and is extremely sensitive to air and water pollution. The biggest threats come from Russian nickel, copper and other industrial plants, acid rain and local farming, logging and pulp processing. The unemployment rate in the region is over 24%, compared with 20% for Finland as a whole. The bulk of employment is in the public sector and in primary production, especially forestry, dairy farming and the paper and pulp industry. Tourism is important in some localities.

A number of general themes run through the programme set out in the SPD (Commission, 1995b). These include

- identification and development of the region's assets and opening up of new areas of activity that will support Finland's position in the world economy;
- promoting internationalisation and co-operation, and seeking every practical way of compensating for the remote location and distance from decision-making centres;

- giving special attention to employment and cultural activities for women and young people in rural areas in order to maintain stability of rural settlements;
- supporting existing agriculture and farm diversification; and
- protecting the fragile natural environment and following the principles of sustainable development.

The programme of proposed actions is grouped under three priorities:

1. Business development and company competitiveness.
2. Development of human resources and expertise.
3. Agriculture, forestry, fisheries, rural development and the environment.

Operational objectives of reduction in unemployment, more private sector jobs and convergence with the national average GDP per capita have been set in terms that reflect the aims of the white paper on *Growth, Competitiveness and Employment* (Commission, 1993b), and especially those of employment-based growth and the opportunities offered by information technology.

Under the first priority, there are seven specific measures designed to stimulate development and diversification of business activity. Creation of an enterprise culture to encourage growth of SMEs is a major theme, as is training and business support services. Telecommunications and renewable energy from the abundant forestry resources are also emphasised.

The second priority closely supports the business and job-creation objectives of the first by the development of research, education, skill training and technology-transfer facilities. The first measure is an ERDF-funded investment programme for higher education to increase the numbers at the two universities in the Objective-6 area, Joensuu and Rovaniemi, and to develop teaching and research supportive of the economic development strategy. Other measures are funded by the ESF and include training and skill development programmes, networking to share and disseminate expertise, and integration into the labour market of young people and those hitherto excluded owing to lack of up-to-date skills or long-term unemployment. The final measure under the second priority is a programme of training, business advice and infrastructure development aimed at the development of the information society and the promotion of distance working.

Much of the programme under the third priority consists of agricultural training and rural development measures and compensatory allowances for farmers typical of Objective 5(a) in other member-states and funded by the EAGGF. All the Objective-6 area has been classified as a less favoured area under Directive EEC/75/268 (see Chapter 5). Under the rural development heading, green tourism such as crosscountry skiing and nature observation is a sector to be developed. Tourism is seen as having considerable potential to attract more non-Finnish visitors to the region, especially from Russia. Tourism, along with much of the rest of the economic base, is dependent on preserving the quality of the environment, and the final measure under the third priority is concerned with its management and protection. The Finnish planning system already gives special protection to shorelines, and these are emphasised here along with waste management and recycling, environmental management plans and land-use plans.

Finally, 40% of the Finnish allocation of Community Initiatives funding is to be within the Objective-6 area, and the SPD indicates the intention to make use of the INTERREG, URBAN and other Community Initiatives (see below). INTERREG will be the most important, and the programme for the Russian border area is outlined in Chapter 8.

CIs

Community Initiatives (CIs) are special financial instruments which the Commission proposes to member-states on its own initiative, designed to help solve problems which it has identified as having a particular impact at the European scale. In a sense, they are a successor to the Community Programmes which formed part of the ERDF prior to the 1988 reforms (Chapter 5). A total of 9% of the Structural Funds is available to support CIs.

The Commission sought to give CIs three distinctive features: a transnational and inter-regional character, bottom-up implementation, and the opportunity to give the EU a high profile in the localities concerned.

In 1993 the Commission proposed that CIs should be based on five priorities:

1. Crossborder, transnational and inter-regional co-operation and networks.
2. Rural development.
3. The most remote regions.
4. Employment and the development of human resources.
5. Management of industrial change.

In order to reflect concerns expressed, especially by the EP, during the consultation process, two further priorities were added:

6. Development of crisis-hit urban areas.
7. Restructuring of the fishing industry.

A total of 13 CIs were adopted by the Commission in 1994. These are listed in Figure 7.7 (Commission, 1994c). Later in 1994, guidelines for their Operational Programmes were published in the *OJ*, and member-states were invited to submit proposals for inclusion.

An outline of these CIs is given here. Several are discussed at greater length in other chapters. The ingenuity of those who devise acronyms for the Commission is evident once again. Reserve, however, is not an acronym. It really is a reserve of about 12% of the available funding to allow the Commission to respond to the experience gained from implementation and other developments or opportunities.

INTERREG II is a combination of two earlier CIs, adopted in 1990, INTERREG and REGEN. Its aim is to promote crossborder co-operation, help isolated areas on internal and external EU frontiers, especially in Objective-1 areas, and fill gaps in energy networks (see Chapters 8 and 9). Over 75% of the

	becus	*Objective 1*
INTERREG II (INTERREG and REGEN)	2.90	2.3
LEADER	1.40	0.9
REGIS	0.60	0.6
Employment (NOW, HORIZON, YOUTHSTART)	1.40	0.8
ADAPT	1.40	0.4
RECHAR	0.40	0.1
RESIDER	0.50	0.1
KONVER	0.50	0.2
RETEX	0.50	0.4
Textiles and Clothing in Portugal	0.40	0.4
SMEs	1.00	0.8
URBAN	0.60	0.4
PESCA	0.25	0.1
Reserve	1.60	0.8
Total	13.45	8.3

Notes
Becu is billion ecus at 1994 prices.
Objective 1 is the amount in becus designated for Objective-1 areas.

Figure 7.7 *List of Community Initiatives*

budget must be spent in Objective-1 regions, and all but a small proportion of the remainder in Objective 2 and 5(b) regions.

LEADER dates back to 1991, so the 1994–9 CI is known as LEADER II. It is designed to stimulate rural development by helping rural associations, local authorities and rural action groups devise innovative strategies for local development, and by helping local people acquire the skills necessary to establish integrated strategies for development based on realisation of the local potential of the areas concerned. It also includes a networking component, to encourage sharing of experience between organisations in at least two member-states, because some of the locally based initiatives it supports are totally new to some parts of the EU (Commission, 1994a: 121–2). Rural localities in Objective 1 and 5(b) areas, plus some contiguous areas, are eligible.

The aim of REGIS is to foster closer integration of the most remote island regions with the EU through economic development, transport and telecommunication links, and vocational training. The eligible areas are the French overseas *départements* (Guadeloupe, French Guiana, Martinique, Réunion), the Azores and Madeira (Portugal) and the Canary Islands (Spain).

'Employment and the Development of Human Resources' is the full title of a Community Initiative established on the basis of the Commission's white paper on *Growth, Competitiveness and Employment*. It incorporates the NOW and HORIZON CIs which were launched in 1990, together with YOUTHSTART which was new in 1994, to create three strands: EMPLOYMENT-NOW, EMPLOYMENT-HORIZON and EMPLOYMENT-YOUTHSTART. EMPLOYMENT-NOW focuses on equal opportunities for women. EMPLOYMENT-HORIZON seeks to facilitate access to the employment market for the handicapped, migrants, homeless, former convicts and other disadvantaged groups, and EMPLOYMENT-YOUTHSTART is intended to encourage the integration into the labour market of people under the age of 20, especially those without adequate qualifications. The programme is available throughout the EU but a substantial priority is given to Objective-1 regions, where over half the budget must be spent. YOUTHSTART is intended to be closely integrated with other ESF measures and with Leonardo (see Appendix II).

The purpose of ADAPT is to assist workers faced with changes in working practices and skill requirements, and support training programmes.

RECHAR dates back to 1989, and is a programme to support the economic conversion of coal-mining areas through the renovation of community infrastructure and the promotion of new economic activities and training programmes. RECHAR II is a continuation, to which areas affected by the decline in lignite extraction (mostly in the former GDR) have been added. To be eligible, areas must normally be under Objectives 1, 2 or 5(b) and have lost or have at risk at least 1,000 jobs in the industry. In addition to the Structural Funds, loans from the EIB and ECSC are available and interest-rate subsidy may be funded.

RESIDER is a very similar programme for areas affected by the decline of the steel industry, and also dates from 1989.

KONVER provides support for economic diversification in areas traditionally dependent on defence industries, through the conversion of the economic base to reduce such dependency and encourage the development from this base of commercially viable activities in all industrial sectors, with the exception of those that might have a military application. Areas eligible must have lost or have at risk at least 1,000 jobs in the defence sector. They need not be in Objective 1, 2 or 5(b) areas although half the funding must go to such areas.

RETEX is a similar programme for areas affected by the decline of the textile industry, to reduce dependence on textiles and at the same time improve the viability of the industry. Most designated areas are NUTS-3 areas eligible under Objectives 1, 2 and 5(b) with at least 2,000 jobs in textiles and clothing.

Textiles and Clothing in Portugal is a specific and exceptional form of supplementary provision, similar to RETEX, for Portugal. It is integrated with Portugal's CSF.

SMEs is another Community Initiative which seeks to translate the ideas of the white paper on *Growth, Competitiveness and Employment* into practice. The purpose is to assist SMEs in any sector to become competitive in the SEM

Rank	Region	Member-state	% GDP per capita 1989–91 (EU 12 = 100)
1	Thüringen	D	30.0
2	Mecklenburg-Vorpommern	D	33.0
3	Sachsen	D	33.0
4	Alentejo	P	33.0
5	Sachsen-Anhalt	D	35.0
6	Voreio Aigaio	GR	35.2
7	Brandenburg	D	36.0
8	Ipeiros	GR	36.2
9	Guadeloupe	F	39.0
10	Centro	P	39.6
11	Dytiki Ellada	GR	40.8
12	Anatoliki Make, Thraki	GR	43.3
13	Ionia Nisia	GR	43.7
14	Thessalia	GR	43.7
15	Réunion	F	45.0
16	Kriti	GR	45.5
17	Kentriki Makedonia	GR	46.8
18	Peloponnisos	GR	47.3
19	Algarve	P	47.9
20	Extremadura	E	49.5
21	Dytiki Makedonia	GR	50.2
22	Norte	P	50.2
23	Notio Aigaio	GR	52.2
24	Attili	GR	52.3
25	Martinique	F	53.0
26	Guyane	F	54.0
27	Andalucia	E	57.8
28	Calabria	I	57.9
29	Sterea Ellada	GR	58.0
30	Galicia	E	58.3
31	Castilla-La Mancha	E	63.1
32	Ceuta y Melilla	E	63.6
33	Basilicata	I	64.5
34	Castilla-León	E	66.7
35	Sicilia	I	67.5
36	Ireland	IRL	68.0
37	Campania	I	70.2
38	Murcia	E	71.3
39	Asturias	E	71.5
40	Puglia	I	74.1
41	Sardegna	I	74.2

Figure 7.8 *NUTS-2 regions ranked by level of GDP per head*

42	Cantabria	E	74.4
43	Canarias	E	74.5
44	Northern Ireland	UK	75.1
45	Comunidad Valenciana	E	76.0
46	Lisboa e vale do Tejo	P	76.6
47	Merseyside	UK	76.7
48	Highlands, Islands	UK	76.9
49	South Yorkshire	UK	77.5
50	Hainault	B	77.6
51	Flevoland	NL	78.1
52	Molise	I	78.8
53	Corse	F	79.8
54	Cornwall, Devon	UK	80.2
55	Northumberland, Tyne & Wear	UK	80.4
56	Clwyd, Dyfed, Gwynedd, Powys	UK	81.1
57	Lüneburg	D	81.9
58	Namur	B	82.6
59	Lincolnshire	UK	83.1
60	Friesland	NL	83.6
61	Cleveland, Durham	UK	83.8
62	Gwent, Mid Glamorgan	UK	84.5
63	Rioja	E	84.6
64	Luxembourg	B	84.7
65	Aragón	E	84.8
66	Shropshire, Staffordshire	UK	84.8
67	Languedoc-Roussillon	F	85.1
68	Trier	D	86.5
69	Limousin	F	86.8
70	Essex	UK	86.8
71	Hereford, Worcs, Warwicks	UK	87.5
72	Dumfries-Galloway, Strathclyde	UK	88.4
73	Drenthe	NL	88.5
74	Gelderland	NL	88.9
75	Pais Vasco	E	89.1
76	Overijssel	NL	89.7
77	Nord-Pas de Calais	F	89.8
78	Ost for Storebit	DK	90.0
79	Kent	UK	90.2
80	Abruzzi	I	90.2
81	Derbyshire, Nottinghamshire	UK	90.3
82	Poitou-Charentes	F	90.5
83	Lancashire	UK	91.1
84	Bretagne	F	91.4
85	Greater Manchester	UK	91.7
86	Auvergne	F	91.7

Figure 7.8 *continued*

87	West Yorkshire	UK	92.2
88	Koblenz	D	92.4
89	Dorset, Somerset	UK	92.5
90	North Yorkshire	UK	92.6
91	Cataluña	E	92.7
92	Lorraine	F	93.1
93	Weser-Ems	D	94.0
94	Bord-Centr-Fife-Lothian-Tay	UK	94.3
95	Madrid	E	94.4
96	Limburg	NL	94.8
97	West Midlands (county)	UK	95.0
98	Picardie	F	95.2
99	Basse-Normandie	F	95.2
100	Midi-Pyrénées	F	95.3
101	Münster	D	95.4
102	Humberside	UK	95.4
103	Oberpfalz	D	95.6
104	Liège	B	95.8
105	Navarra	E	95.9
106	Niederbayern	D	96.5
107	Schleswig-Holstein	D	96.9
108	Pays de la Loire	F	97.6
109	Hampshire, Isle of Wight	UK	97.9
110	Gießen	D	97.9
111	Baleares	E	98.3
112	Noord-Brabant	NL	98.6
113	Bourgogne	F	98.9
114	Umbria	I	98.9
115	Surrey, East & West Sussex	UK	99.4
116	Vest for Storebit	DK	99.4
117	Unterfranken	D	99.6
118	Oost-Vlaanderen	B	99.7
119	East Anglia	UK	99.8
120	Provence, Alpes, Côte d'Azur	F	101.5
121	Bedford, Hertfordshire	UK	102.6
122	Centre	F	103.0
123	Cumbria	UK	103.3
124	Limburg	B	103.3
125	Acquitaine	F	103.3
126	Oberfranken	D	103.9
127	Cheshire	UK	104.0
128	Arnsberg	D	104.2
129	Franche-Compté	F	104.2
130	Leicester, Northampton	UK	104.6
131	Marche	I	104.7

132	Detmold	D	106.0
133	West-Vlaanderen	B	106.3
134	Zuid-Holland	NL	106.3
135	Avon, Gloucester, Wiltshire	UK	106.6
136	Utrecht	NL	107.1
137	Kassel	D	107.4
138	Saarland	D	107.6
139	Haute-Normandie	F	108.4
140	Zeeland	NL	108.7
141	Toscana	I	109.4
142	Rhône-Alpes	F	109.7
143	Freiburg	D	110.0
144	Champagne-Ardenne	F	110.7
145	Berks, Bucks, Oxfordshire	UK	110.8
146	Rheinhessen-Pfalz	D	110.9
147	Schwaben	D	111.1
148	Braunschweig	D	112.3
149	Köln	D	112.8
150	Tübingen	D	112.9
151	Alsace	F	113.7
152	Noord-Holland	NL	113.8
153	Liguria	I	115.8
154	Berlin (West)	D	116.3
155	Hannover	D	116.6
156	Brabant	B	116.6
157	Veneto	I	116.6
158	Lazio	I	116.8
159	Grampian	UK	117.3
160	Piemonte	I	119.6
161	Friuli-Venezia Giulia	I	121.6
162	Trentino-Alto Adige	I	122.0
163	Düsseldorf	D	122.8
164	Karlsruhe	D	124.2
165	Antwerpen	B	125.6
166	Hovedstadsregionen	DK	126.7
167	Mittelfranken	D	126.7
168	Luxembourg	L	127.2
169	Groningen	NL	127.4
170	Emilia-Romagna	I	127.5
171	Valle d'Aosta	I	129.6
172	Lombardia	I	134.7
173	Stuttgart	D	137.6
174	Oberbayern	D	148.1
175	Bremen	D	149.7

176	Greater London	UK	151.2
177	Darmstadt	D	162.9
178	Ile de France	F	166.8
179	Hamburg	D	194.5

Figure 7.8 *NUTS-2 regions ranked by level of GDP per head (Note*: NUTS-2 regions, except new *Länder* of Germany, which are NUTS 1, 1991 data. The average of 100 is for the EU of 12, excluding the new *Länder*. *Source*: Commission 1994b, Table A.25, pp 192–4.

and international markets. SMEs are defined for this progamme as employing no more than 250 people, limited turnover and not more than 25% of their capital held by larger companies. Of the budget, 80% must go to Objective-1 regions, the rest to Objectives 2 and 5(b). Funding may be linked to EIB and EIF loans, for which it may include interest rebates.

URBAN is intended to extend and improve co-ordination of EU measures directed at urban problems, and is discussed in Chapter 11.

PESCA is specifically targeted at the crisis of overcapacity in the fishing industry by assisting worst affected areas. All but 15% of the budget is allocated to areas in Objectives 1, 2 and 5(b).

Periodic reports and overviews

DG XVI is required to produce periodic reports on the social and economic situation and development of the regions in the Community and on the operation of the Structural Funds. These provide extensive sources of data and commentary on the extent of disparities between the different regions of the EU, including league tables of relative prosperity. The first periodic report, *On the Social and Economic Situation in the Regions of the Community*, was published in 1980 (Commission, 1980a).

The fifth periodic report, *Competitiveness and Cohesion Trends in the Regions* (Commission, 1994b), was published to coincide with the 1994–9 programme period and enlargement and is a valuable source of statistical information, as were the earlier periodic reports. Some of its data necessarily relate to the EU of 12, and some to the anticipated EU of 16. A full set of data of this type for the existing EU of 15 member-states is awaited in the next periodic report. For those who like to see where they are in league tables, these periodic reports offer plenty of scope (see Figure 7.8). Each of the NUTS-2 regions is ranked according to its GDP.

The funding and incentives offered by the EU Structural Funds is, for most of the regions within objective areas, only one of the regional policy measures and sources of development funds that may be available. National and regional authorities often also offer development incentives. The most comprehensive listing for the whole EU is Yuill *et al.* (1995).

It is not the purpose of this book to make a critical evaluation of what the Structural Funds achieve, as there are several sources of further reading on regional policy and economic development. See, for example, Alden and Boland (1996), Chisholm (1995), Cole and Cole (1993), du Granrut (1995), Hardy *et al.* (1995), Vickerman (1992).

8

Hands across the border: networking, lobbying and crossborder planning

Interaction between authorities and organisations from different member-states has become a major feature of professional life throughout the EU. This chapter concentrates on three ways in which this takes place. Of particular concern to spatial planning authorities are networking, lobbying and crossborder planning.

The first part focuses on non-tangible forms of networking rather than those involving physical links. This is in a sense the distinction between networks, in other words, between those that are communication networks and associations, and those that involve physical infrastructure. The latter, in the form of TENs, are the subject-matter of Chapter 9. The former often seem intangible, being more dependent on exchange of information than of people or goods, but they are an important element of EU policy-making.

Many networks count lobbying among their objectives, and lobbying itself is a very prominent feature of EU activities in Brussels. It is estimated that as many as 10,000 people are employed as lobbyists in around 3,000 firms or representative offices in Brussels. The lobby and lobby networks are discussed in the second part of the chapter.

In some locations, such as the Dutch border regions and around Aachen, crossborder co-operation in planning has long been accepted as a normal part of spatial planning and policy-making by the local authorities. It is important to emphasise the difference between this form of local planning, which is quite orthodox in the localities concerned, and crossborder initiatives dating back to the 1980s or 1990s, which represent a break with conventional planning parameters and are more truly a response to the European context of spatial policy-making.

Several crossborder planning initiatives, especially those that are more innovative, are supported by the same programmes (such as RECITE and INTERREG) that have supported the growth of networking and lobbying organisations. It is appropriate therefore to include this aspect of planning here as the third part of the chapter.

Networking

Possibly as a result of the intangible nature of some networking activities, the topic of city networking has received much less academic attention than it deserves. Nevertheless, city networking has become a highly significant and widespread phenomenon which has grown rapidly in western Europe with the establishment of associations such as Eurocities and the Union of Baltic Cities, and EU programmes such as RECITE, Ouverture and ECOS.

In the sense of intangible networking under discussion here, the term includes the development of all forms of linking, associations, twinning and other relationships between cities (or between regions or other types of organisations). In general, networking in this sense does not in the first place imply links in the form of physical infrastructure or transport connections, although the promotion of transport and telecommunications networks may follow from contact networking of the sort meant here. They can operate satisfactorily with the infrastructure that already exists, and often without recourse to recent advances in information technology, although the need for improvements to existing transport infrastructure may often be identified.

In spite of the possibilities offered by telecommunications and information technology, the growth of networking probably tends to increase the demand for travel, rather than substitute for it, since sooner or later many participants feel the need for personal contact or the opportunity to see for themselves the cities and regions in which their partners are situated, the planning problems they face and the development projects undertaken there. Travel throughout the EU is on the increase and networking certainly contributes its share of this growth, and it probably remains the case that the transfer of information in the heads of people who travel from city to city continues to be the highest-capacity method of transferring information around Europe.

City networking has received disproportionately little attention from academic researchers in comparison with the level of activity (but see Dawson, 1992; Nijkamp, 1993; Parkinson, 1992; Robson, 1992; Stoker and Young, 1993).

Purposes of networks

Several different forms of networking and categories of networks can be identified. Classification of networks is not easy because their objectives and activities do not fall into neat categories. There are a number of dimensions which apply to networks whose activities relate partially or wholly to spatial planning and policy-making.

The variety to be found among these networks is indicated by the following dimensions:

- Whether the objective of the network is political or economic, for example in connection with lobbying, trade promotion, city marketing, technology transfer.
- Whether the members of the network choose to participate because they have some basic similarity, for example a similar traditional industrial base, or whether the attraction is to link with places in contrasting circumstances.
- Whether the network is for the purpose of communication of ideas and information, of people or of goods.
- Whether the objectives are more rhetorical or practical.

- Whether the network is constituted or incorporated as a formal association, or whether it is informal, based on personal contacts without any institutional basis.
- Linked to the above, whether the network is a permanent or previously existing organisation or whether it is an *ad hoc* body formed opportunistically in order to take advantage of an EU programme such as RECITE.

Stoker and Young (1993) suggest that networking represents a whole new mode of operation for local authorities, contrasting to their traditional hierarchical style and the marketing and city imaging of the 1980s. The question posed is how can informal and intangible networks be moulded into a purposeful framework for strategic action, justifying the time, effort and expense involved in creating and sustaining them?

New associations or representative bodies tend not merely to respond to the existing agenda and undertake the tasks for which they were first set up; they frequently have the effect of taking this agenda further. This may be partly explained by such organisations feeling the need to justify their existence and budgets by demonstrating the range of tasks they should undertake. However, it is also a consequence of the stimulus resulting from a coming together of ideas and experience from different ways of thinking, different cultures and different parts of Europe.

Networks are being developed to represent common interests (for example, of automobile-manufacturing cities, former coal and steel cities, etc.) with a lobbying role in relation to the Structural Funds. There are mutual support networks, assisting, for example, with business contacts; networks concerned with policy development seeking to influence policy-making in the Commission; technology transfer networks especially in relation to CEE countries (e.g. the Ouverture programme); and networks of academic and professional associations, such as AESOP.

The advent of the SEM has led to a growing sense of competition between cities and regions throughout Europe as a whole. As the then Italian Foreign Minister, Gianni de Michaelis, said during his country's presidency in 1990, 'the single market is not so much the competition between companies as the competition between systems, and those countries and regions with the most appropriate systems will be the winners'. Networking is a manifestation of this growing sense of competition between cities in Europe, along with the phenomena of city imaging and marketing and the concept of spatial positioning (see Chapter 6).

Why join?

From the point of view of individual municipalities, the case for participating in such networks is not proven, but the attitude often is that membership is likely to offer advantages in the form of information, access to expertise and a role in shaping future EU policy, so it would be disadvantageous to opt out.

Joining a network may not be difficult, and the cost of a subscription quite modest. But it is not always enough simply to join such a club. Member cities often take the view that the important point is to get into a position of influence on some committee or working party concerned with a sector of policy. In this way, they may put themselves into a position to influence any position papers, lobbying or advocacy undertaken in the name of the network or association (Mazey and Richardson, 1993). This way, a city may have a role in agenda-setting on issues which are also on the agendas of the EU Commission and national governments, giving it influence or at least advance information in the EU policy community. Participation in this way requires a commitment of resources that may far exceed any subscription charge, in the form of the time and travel expenses required to attend meetings and to draft reports.

Spatial policy networks

There are not many networks that are exclusively concerned with spatial planning although one concerned with spatial research is intended to be created in 1996 by the Committee of Spatial Development at the request of the Informal Council of Ministers (see below). The major reason why so many local and regional authorities devote so much time and effort to networking within the EU is that so many EU policies and programmes impact directly on them either as a context in which their own policy must be formulated or as the agency of implementation. Therefore, networking arises out of the whole range of local and regional responsibilities, or be generated by any sector of them.

So, although not so many are narrowly focused on spatial planning, a very large proportion of networks and associations would emphasise spatial policy among their concerns along with economic development, transport, environmental planning and urban policy. A number of influential examples of networks and programmes that address spatial themes are described in more detail below. There is wide support within local government for the suggestion that, at the forthcoming IGC, revisions to the treaty should include giving formal competence to the EU over urban policy. Many of the networks and associations referred to here are active in this campaign, which is discussed further in Chapter 11.

Two important bodies that are directly concerned with spatial planning are the European Council of Town Planners (ECTP) and AESOP. These are both quite specialist, being concerned with professional expertise and education in planning rather than spatial policy as such, and are discussed in Chapter 13. Networking figures among their informal objectives, especially for AESOP, which has been the catalyst for many personal and interuniversity links.

Origins and funding

A number of organisations that are involved in European networking have been in existence for many years, but there was a considerable growth in interest and in the number of initiatives taken during the 1980s to set up new European associations. For example, the ECTP was founded in 1985, Eurocities in 1986 and AESOP in 1987. Alongside specific motivations which led to the foundation of these and many other associations, the general impetus in the 1980s was provided by the SEM 1992 programme; increasing numbers of funding and other EU programmes in which local and regional authorities could participate; and recognition that effective lobbying of the EU institutions required building coalitions of interest and support from representatives of several member-states (see below).

The Commission generally welcomed the growth of Europe-wide bodies and in several cases gave financial support for specific activities or some degree of recognition as a body to be consulted. Since 1989, a number of these have benefited from direct funding from the Commission under Article 10 of the ERDF regulation, enabling them to become fully established more quickly, to develop more activities or to extend their operations in directions which may otherwise have been difficult – such as inclusion of participants from CEE countries. Commission funding has also supported the foundation of new networks and networking projects.

Article 10

Article 10 provides that the ERDF can provide financial 'support for studies or pilot schemes concerning regional development at Community level, especially where frontier regions of Member States are involved'. Up to a maximum of 1% of the ERDF budget can be allocated under Article 10. The two key categories are studies undertaken on the Commission's initiative and pilot schemes.

Studies may be undertaken to identify

- the spatial consequences of national projects, especially major infrastructure, whose effects extend beyond national boundaries;
- measures which could overcome specific problems of border regions both within and outside the EU; and
- the elements necessary to establish a prospective outline of the utilisation of the EU's territory.

The latter is the heading under which the *Europe 2000* and *Europe 2000+* studies were funded.

Innovative or pilot schemes may be funded which

- constitute incentives for creation of infrastructure, investment in enterprises and other measures meeting EU interests, especially in border regions within and outside the EU; and
- encourage pooling of experience and development of co-operation between different EU regions, and innovative measures.

They key feature of pilot or innovative schemes (the two words are used interchangeably in general EU usage in reference to this category) is that projects should have a high degree of replicability. This example of Brussels English means that projects should be designed to try out and demonstrate ideas which could be widely applied elsewhere.

RECITE

DG XVI has provided funding for direct co-operation between city and regional authorities since 1989 and in 1991 launched the RECITE (Regions and Cities of Europe) programme. This is funded as an innovatory measure under Article 10 and administered on behalf of DG XVI by the RECITE office in Brussels, managed by ECOTEC, a firm of consultants. The RECITE office provides technical assistance and monitors the programme of inter-regional co-operation projects and urban pilot projects (see Chapter 11), publishing reports and newsletters.

RECITE is open to any local or regional authority of over 50,000 population, and groups of authorities and international associations representing cities and regions. The underlying rationale is that, by supporting economic development of participants through the promotion of economic and political linkages, RECITE makes a contribution to the economic and social cohesion of the EU as a whole.

RECITE projects must be concerned with policy sectors that are normally at least partially the responsibility of local and regional authorities. The RECITE programme funded 37 networks in 1990–2 for projects of one to three years' duration. In some cases, the RECITE money was in effect start-up money, and other resources would be required (for example, subscriptions from members) if the network is to continue. In other cases RECITE funded projects or activities within networks that already existed, and in others the project itself may have a limited time. Typically, RECITE contributed 50–60% of the costs of the specific programme of activities approved for funding.

RECITE has supported a number of networks whose concerns are closely linked to spatial policy. Chief among these is Eurocities, discussed below, for which RECITE cofinanced three subprojects: on the use of IT to provide business and visitor information; urban management; and business best-practice advice in anticipation of the SEM. Two others with similar urban policy concerns are the Medium-Sized Cities Commission and Quartiers en Crise. The European Urban Observatory, based in Barcelona, is a database and decision-support system for a network of ten cities (Amsterdam, Athens, Barcelona, Berlin, Birmingham, Brussels, Genoa, Lille, Lisbon and Milan). Ouverture and ECOS (Eastern Europe City Co-operation Scheme) are directed at building links with the CEE. Another that specifies planning and land development among its concerns is EUROSYNET, a group of five authorities whose programme includes recycling of derelict land for business purposes.

The range in size of networks is enormous, from the Medium-Sized Cities Commission with around 100 members to ROC-NORD with only two members from the northern and southern limits of the EU of 12, Crete and north Jutland, to show peripheral solidarity and to transfer technical assistance.

The ingenuity of some of the acronyms of the Community Initiatives (see Chapter 7) is matched by those of some RECITE projects and networks, for example CAR (Co-operation between Automobile Regions) with eight members and COAST (Co-ordinated Action for Seaside Towns) with nine members. The parallels with the Community Initiative programmes and the Structural Fund objectives can extend further, for example one network of nine members explicitly calls itself 'Economic Development of Less Favoured Regions', while 'Co-operation between Atlantic Regions' consists of 16 members concerned with problems of peripherality. Greater extremes of peripherality are found in EURISLES, a project for islands within the EU launched by a longer-standing representative body, the Conference of Peripheral Maritime Regions of the EU (CPMR). The CPMR has its offices in Rennes in western France, and EURISLES includes non-European parts of the EU (Réunion and Martinique) as well as Atlantic islands (the Azores, the Canaries and Madeira), Corsica, the northern Aegean islands and the rather less peripheral Isle of Wight.

The link with the Community Initiative programme is most explicit with a network called DEMILITARISED. DEMILITARISED is an acronym standing for 'Decrease in Europe of Military Investment, Logistics and Infrastructures and the Tracing of Alternative Regional Initiatives to Sustain Economic Develoment'. It brings together 16 authorities whose economy has been heavily dependent on military facilities, and who have consequently suffered from the disarmament of the 1990s. The network proposed urban and regional policies for economic and infrastructure development, environmental protection, housing, social and cultural development – through which they sought to achieve higher public awareness of their problems and direct support from the EU through a funding programme.

The KONVER Community Initiative (see Chapter 7) is, at least indirectly, the outcome of lobbying by DEMILITARISED. KONVER started in 1993 as an annual programme based on an earlier special programme called PERIFRA which operated in 1991–2 and was itself introduced in response to calls from the EP. KONVER now runs until 1997 and comes within the 1994–9 programme of Community Initiatives.

Eurocities

Eurocities is a good example of a network of cities now becoming very active, having opened an office and secretariat in Brussels with help from DG XVI early in 1992. A wide range of crossnational policy working parties and commissions have been set up, covering all aspects of the range of responsibilities held by municipalities. Consequently, Eurocities has links with 14 of the 24 DGs. Much of its work is on spatial and urban policy issues such as economic development or the urban environment. Eurocities was also responsible for coordinating the launch of the Car-Free Cities Club in Amsterdam in March 1994 (see Chapter 11).

Eurocities is in the position of being selective of who it admits, and applicants have been turned down. Members must be major cities of at least 250,000 population which have a democratically elected city government, must be important regional centres and must play an international role. In early 1995, there was a total of 52 full members (who must be from EU member-states) and 10 associate members (from non-EU European countries). These include most of the major cities in the EU, with the exception of four of the national capitals, Brussels, London, Paris and Rome. This partly reflects its origin as a network of second cities initiated by Barcelona, Birmingham, Frankfurt, Lyon and Rotterdam. London, however, is not eligible as it has no elected city-wide government.

Telecities

Much of this networking does not imply any physical link in the form of transport infrastructure, nor any modes of communication beyond those conventionally available.

However, one of the commissions within Eurocities, Telecities, is seeking to develop and promote the use of telecommunications between city administrations (initially between its own members). This is in order to create a sense of closer linkage through immediacy of communication and thereby make the transition from being a network to a policy community – with the greater cohesion and lobbying power this implies.

Telecities is concentrating on the application of telematics, i.e. telecommunications and information technologies, to support economic and social regeneration of urban areas, including the use of teleworking within employment strategies. Telecities is funded by DG XIII and is co-ordinated by Manchester as lead city, together with other members of the steering committee, Antwerp, Barcelona, Bologna, Den Haag and Nice.

It is argued that in the 1980s there was a perceived technological gap between the EU and both the USA and the advanced countries of the far east in terms of telematics, and therefore more research and adoption of new technology was needed by the EU. This phase created a situation where the technical capability of telecommunication systems exceeded the understanding of the policy community, who hence did not use this technology. The Telecities group is placing the emphasis on user orientation, shifting from technology push to demand or end-user pull, so that telecommunication systems can be developed in ways that facilitate the exchanges and serve the communication needs city networking generates. The technology therefore aids the process of building policy networks at the EU scale.

RETI

The Association of European Regions of Industrial Technology (RETI, after the title in French, Régions Européennes de Technologie Industrielle) is less comprehensive in its membership, but does represent a large number of regions dependent on traditional heavy industries such as coal, steel and shipping. It was founded in 1984 in Lille by four regions, Hainault (Belgium), Nordrhein–Westfalen (Germany), Nord-Pas de Calais (France) and West Yorkshire (UK). Its main objective is to provide a platform for industrial regions to define their place within the framework of EU policy, especially the Structural Funds, and to present their specific problems in a unified manner.

RETI now has over 20 members, mostly drawn from regions concerned with ECSC programmes and the Structural Funds, especially Objective-2 regions. It includes by no means all potential members, drawing only from Belgium, France (e.g. Lorraine), Germany (e.g. Saarland and Sachsen–Anhalt but no longer the Ruhrgebiet), Italy (e.g. Toscana), Spain (e.g. Asturias), The Netherlands (e.g. Zuid Limburg) and the UK (most of the traditional coal and steel areas except the north-east of England).

Quartiers en Crise

Quartiers en Crise is the name of a network of over 25 cities which all have experienced serious urban decline with the attendant problems of poor housing conditions, unemployment, crime and other inner-city social problems. The network began with 10 members in five countries as an exchange programme for urban planning professionals working in deprived areas. The network developed its own approach to an integrated community-based strategy for such areas, which attracted interest from many other cities with similar problems. Within the framework of the RECITE programme, Quartiers en Crise (it has not adopted an English title) sought to develop its co-operation and exchange programmes – and its ideas for an integrated response at both national and EU levels – to urban decline through research, training and the transfer of knowledge and experience, especially between northern and southern member-states. It is probably not coincidental that no fewer than 12 of the cities selected for Urban Pilot Projects (see Chapter 11) are members of Quartiers en Crise.

Medium-Sized Cities Commission

This is a grouping of over 100 cities, many of which are also members of Eurocities. It performs a similar function but accepts a wider range of cities into membership, a population of 50,000 being the threshold. Within the RECITE programme, it undertook four specific co-operation projects: on training for the young unemployed; the urban application of computer technology; guidelines on science and technology park development; and urban renewal by means of non-polluting small enterprises suitable for the central areas of tourist cities.

Ouverture and ECOS

Ouverture is a programme funded under Article 10 of the ERDF regulations and is co-ordinated from Strathclyde Regional Council in Glasgow. It was launched in 1991 and, with a budget of 5 mecus was the biggest single Article-10 project. Its purpose is to support the development of democratic subnational government and assist municipalities of former communist countries in the CEE to develop networking links and relationships with municipalities in EU member-states in order to promote advice, exchanges, business and trading links, transfer of technical expertise, etc.

Ouverture is itself a product of networking. It is the initiative of five regions, Asturias in Spain, Piemonte in Italy, Saarland in Germany and Midi-Pyrénées in France, as well as Strathclyde, who had themselves come together as a network to share their experience and offer mutual support to overcome their economic development problems.

Until 1995, Ouverture has run in parallel with ECOS, a scheme within PHARE to facilitate co-operation between local and regional authorities both within the EU and between the EU and the CEE. ECOS started in 1992 under the management of the Council of European Municipalities and Regions (CEMR) and the City of Strasbourg. ECOS has a more land-use planning and urban policy orientation, whereas Ouverture has a more economic and regional development emphasis, but there is considerable overlap in their respective scope. For this reason, they merged in October 1995. The headquarters remain in Glasgow with offices in Brussels and Paris. The merged Ouverture will have a budget of around 12 mecus p.a.

Ouverture sees itself as an element of spatial policy embracing the EU and CEE, with an emphasis on peripheral and disadvantaged areas in both groupings. From the point of view of peripheral regions in the west of the EU, such as Strathclyde itself, Ouverture provides local businesses with the opportunity to compensate for the eastward shift of the European centre of gravity. Local and regional authorities are invited to propose projects for funding, usually of up to one year's duration. Projects must involve at least two authorities from two different EU member-states plus at least one from a country eligible under the PHARE programme, or since December 1993 the TACIS programme, so that the CEE partner has at least two models of western practice to learn from. Although they must be led by local or regional authorities, an Ouverture project can include businesses, consultants and universities as partners or advisers.

Projects usually include a combination of the following types of action:

- Exchange of experience and know-how.
- Training and study visits, including the production of training material.
- Secondment of technicians and experts, staff placements, job-shadowing and staff exchanges.
- Provision of advice and expertise.
- Participation in, and organisation of, seminars, trade fairs, exhibitions and company visits.

Approximately 250 projects involving over 1,000 local and regional authorities had been funded up to 1995.

In assessing proposals, priority on the EU side is given to partners from Objective 1, 2 and 5(b) areas, and about 80% do have such partners. Grants of 60% of costs are available, except in Objective-1 areas where 75% is payable. Highest pro rata participation rates are found in the EU periphery of Greece and Ireland. On the CEE side, preference is also given to peripheral and less favoured locations. Visegrad countries have predominated, using Ouverture as part of their preaccession strategy. This will change as they become more integrated into the EU economy, and realise the advantages of their central Europe location while more links and projects are developed by the CIS countries.

Baltic Gateways

To illustrate the Ouverture programme, reference can be made to the Baltic Gateways project. This began in 1991 and is co-ordinated by North Tyneside, a municipality adjacent to Newcastle upon Tyne (Williams, 1993a; 1993b). The aim is to link North Tyneside with Esbjerg in Denmark, Rostock in Mecklenburg–Vorpommern (one of the new *Länder* of Germany), Gdynia in

Poland and, in 1993, Kleipeda in Lithuania. The aim is to meet the objectives of Ouverture by generating a programme of technical assistance and mutual support, and in so doing to revive some of the traditional trading routes among Tyneside, the Baltic and the Hanseatic ports. Thus, it fits within the Northern Arc concept (see Chapter 6).

A preliminary round of visits took place to identify themes and develop issues to be pursued. The issues identified included: city administrative structures for economic development; port authority infrastructure development and relationship with the city council; industrial restructuring and the impact of new industries; and tourism and economic development. A second round of meetings was held in 1992–3 to deepen the understanding of these issues as they affected each partner city. These meetings included a number of speakers representing the various agencies and authorities operating in each area, plus detailed guided tours of the port facilities.

The conclusion of the project team itself is that very worthwhile initiatives have been taken and links established (CLES, 1992; Scott, 1993b). Early meetings were essentially opportunities to get to know one another, understand each other's level of awareness, technical and information resources, and appreciate the physical and institutional context of those responsible for planning and economic development in each of the participating port cities.

They identified a number of specific ways in which advice could be offered, expertise shared or transferred, or mutual benefit achieved from co-operation. An example was the case of a UK furniture manufacturer who imported chipboard from southern Poland. Neither the supplier nor the importer were committed to any particular Polish and UK ports. The Baltic Gateway network promoted the use of Gydinia and the Port of Tyne for this trade.

The main outcome, however, was a mutual learning experience. It took some time to reach a point where realistic and worthwhile proposals could be made regarding specific exchange, technology transfer or other interaction.

Participants observed that people new to this form of international interaction, especially from CEE countries, have gone through phases of simply being delighted to participate and to learn about anything they can, and are moving towards a position of developing very clear ideas of why they are in and what their agenda is. This was very evident in the case of Polish participants, an indication of the success of the network. The timescale was possibly rather short for some others to undergo the learning process necessary to take full advantage of the opportunities offered from the contacts and other EU programmes.

For the delegations from North Tyneside and Esbjerg, the learning experience focused mostly on appreciating the reality of the situation in the former communist cities not only in the categories where problems were to be expected: environment, physical infrastructure, institutions and financial resources; but also it became clear that it was necessary, but not always easy to achieve, an appreciation of what and how much the municipal officials did not know, what assumptions to make and where to begin in order to avoid the twin dangers of either patronising or confusing them. The western cities also had two other tasks were easy to underestimate: that of convincing the former communist participants that they do not have access to unlimited wealth and educating them about the limitations of their powers in both the financial and legal senses.

Spatial research networks

The idea of establishing a Co-operation Network of Spatial Research Institutes was developed in 1993 by the Committee of Spatial Development (CSD) and DG XVI, and strongly promoted by the 1994 German presidency. The term 'observatory' is sometimes employed, implying a research institute with its own scientific staff, but it is more likely that the network model will prevail. In essence, it will take the form of a network of academic and research institutes nominated from each member-state, who will form an advisory group and body of expertise that the Commission and the CSD can call upon. Members would in turn call upon the expertise of other universities and institutes from their own countries. It was originally anticipated that it should come into operation during the Spanish presidency in 1995, ready to assist with the preparation of the *European Spatial Development Perspective* (see Chapter 12), but it has been delayed.

In addition to being the driving force behind the observatory or co-operation network concept, the German government has itself sponsored a network of spatial research institutes in the CEE as part of its 'Program of cooperation with countries of Central and Eastern Europe and of the Commonwealth of Independent States in the field of spatial planning' (BMBau, 1995b: cover page). Other elements of this programme are the elaboration of a crossborder spatial development concept for the Germany–Poland border region (see below), assistance for CEE countries that are participating in the vision and strategies around the *Baltic 2010* project (see Chapter 6), and a programme of technical and legal advice and study visits for people responsible for urban planning, housing policy, economic promotion and construction in CEE countries. This programme is to help adjust not only to the market economy but also to EU environmental and construction standards.

Lobby and policy

Many networks include lobbying, or seeking to influence EU policy or its implementation, among their objectives. Some of the larger networks, such as Eurocities and RETI, maintain an office in Brussels for this purpose. The estimated 3,000 organisations that form the lobby in Brussels include many other categories. Many are industry or trade associations, or represent the interests of different professions, but representation of local and regional authorities, or groupings of them, forms an important part of the lobby.

Several of the networks discussed above have developed to the extent that they now feel that maintaining an office with full-time staff in Brussels is both financially feasible and justified in terms of achieving their objective.

Any organisation that seeks to be accepted as a recognised voice of a group of cities or regions with a common interest, a professional or trade association, or in fact any non-government organisation, needs to be able to demonstrate its standing to the Commission. This normally requires demonstrating that it does represent a broad membership drawn from several member-states, and that it is acknowledged to be the recognised body. This is especially important for any trade or professional body, but for networks in programmes such as RECITE it does not matter that similar objectives are pursued by more than one network. This is better than loosing cohesion through becoming unmanageably large.

Lobbying is not only directed at the Commission, of course. Lobbying may also be directed at committees of the EP, CoR and Ecosoc. Parliament is such a major target that many organisations install a team of representatives in Strasbourg for the weeks in which the monthly plenary sessions occur there. Proposals to set up an official register of lobbyists, allowing only those who register to have privileged access to MEPs inside the building, were approved by the EP Rules of Procedure Committee in 1995 (EIS, 1995: 29) and will be submitted for approval by a plenary session of the EPO during 1996. Handling the lobby was seen as one of the key issues to be addressed when working arrangements were first made for the new CoR, who feared that members would be 'bogged down by lobbying activities' (EIS, 1994a).

One of the unwritten rules which motivates all networks that seek to obtain EU funds for projects, or to be recognised as representative of certain interests, is that the network must contain members from as many member-states as possible, and certainly over half, unless the common interest represented (such as viticulture) can only occur in certain geographical areas. Otherwise, it always helps if membership is drawn from different parts of the EU including rich and poor, northern and southern regions.

The Commission and the lobby

Conventionally, it is assumed that networks and lobbyists that maintain an office in Brussels do so in order to lobby, or seek to influence the Commission. For some, the purpose of maintaining a presence is to listen and be able to advise on new political developments or policy proposals coming from the Commission. In reality, the picture can be much more complicated.

From the point of view of the Commission, it is not simply a process of listening or of releasing information. The Commission itself may wish to influence opinion in the EP, CoR or Ecosoc, or in some national governments, and therefore encourages networks and lobby associations to support its case in their own lobbying.

The Commission is conscious of the democratic deficit, and of the pressures to ensure that the concept of a people's Europe is more than mere rhetoric. One way in which it may be able to show that its proposals respond to widely supported views of the population is by being able to point to pressure for such proposals coming from organisations representing local communities and interest groups from different parts of the EU. There is evidence that this is a lesson that DG XI has drawn from the failure to sustain the urban environment initiative following the green paper of 1990 (see Chapters 10 and 11), and that it is actively encouraging networking and lobbying.

Lobby networks can also strengthen the Commission's case whenever it faces resistance from national governments by enabling the Commission to say that its proposals are consistent with the views of local and regional organisations throughout the EU, without being dependent on local opinion as reported to it by national governments. Lobbying by networks can therefore assist in the bypassing-the-capital effect (see Chapter 8), which can sometimes be as welcome to Commission officials as to the network's members.

Representation of subnational government in Brussels

Many local and regional authorities feel the need to establish an office or representation in Brussels. This, like some networking, can be seen as a manifestation of the 'bypassing-the-capital' effect.

The activities these offices undertake may take the form of lobbying, networking, information gathering and promotion of awareness of their locality. In the case of the federal member-states, Austria, Belgium and Germany, these offices also need to undertake tasks which other countries may assign to their national representation or Coreper member because the states are often responsible for legislation on a wide range of sectors of EU policy. Participation from the federal states of these countries in Council of Ministers meetings is often necessary. These need to maintain quite large offices in Brussels. For instance, two of the new *Länder* of Germany, Sachsen and Sachsen–Anhalt, use the building that was once the embassy of the GDR to Belgium.

Member-states where regional authorities do not have such extensive powers may not have such large representations as Germany, while those with a centralised system of government or without any regional tier of government are more likely to rely on local and regional groupings of municipalities and local authority associations for representation of their interests in Brussels.

There are about 70 offices in Brussels representing subnational authorities, all undertaking activities in the form of lobbying, networking, information gathering and promotion of awareness of their locality. They provide a briefing and reporting service to their sponsors and an information service for local officials and politicians, alerting them to opportunities and new policy developments. They also respond to requests from sponsors and local authorities for advice, contacts and help with fact-finding visits and meetings with Commission officials. Staff from these 70 representations in Brussels network among themselves, and this facilitates partner searches for individuals, authorities and companies seeking wider European links in order to undertake joint projects or ventures, or may enable them to advise on the merits of joining some new network or association.

They also often provide an important support service to MEPs and members of CoR from their local area, from whom they will at the same time receive political advice on issues being raised by the EP or CoR. This may assist them to anticipate opportunities under future programmes. Irrespective of whether they are advising MEPs and CoR members, the staff of the local representations in Brussels are available to attend hearings or listen to debates of the CoR and EP committees.

The lobbying function may not be much different from that of promoting awareness of their city or region, but it could take very direct forms. For example, they may seek to ensure that the type of economic or environmental problem that is faced in their locality remains within, or is added to, the scope of the Structural Funds or Community Initiatives so that the authority can take advantage of them. Lobbying involves a lot of person-to-person contact, which is easier to maintain from a local base. Although policy development often seems to proceed very slowly, changes and approvals can occur quickly – so being on the spot can be essential if a quick reaction is necessary.

In the case of the UK, the pattern of local representation is extensive but quite mixed. With a total of 21 such offices, the UK has more subnational offices than any other member-state (John, 1994). Many of them are quite small and share facilities or occupy one or two rooms rented from a host organisation. Another pattern is to hire a consultant who devotes an agreed proportion of time to the authority. The Local Government International Bureau maintains an office which plays host to a number of local authority representations. The Association of London Authorities and the Convention of Scottish Local Authorities also maintain offices there. A few large cities, including Birmingham and Manchester, have their own offices, but it is more usual for a county or a group of counties to combine, often in partnership with training and enterprise councils and other public bodies from the region. In some cases, such as the North of England Office, the whole of a NUTS-1 region is represented, but this is unusual.

However, the objectives of the North of England Office are quite typical and are worth quoting as an illustration:

- to establish a North of England regional presence in a Europe of the Regions;
- to promote awareness of the North of England;
- to increase regional awareness of EU initiatives and policies;
- to influence and inform EU policy-making as it affects the North of England;
- to provide a base for regional organisations on business in Brussels.

(NEA, 1994)

The North of England Office, like many others, is an example of public-private partnership. It is sponsored by the North of England Assembly of Local Authorities, representing all the local authorities in the region, plus the training and enterprise councils and the Northern Development Company whose members include the major firms and development agencies in the region.

Crossborder planning

Co-operation between regions, including crossborder co-operation, is seen by DG XVI as a key factor in the achievement of its cohesion objectives. Consequently, although crossborder planning may be thought of as something rather different from other forms of networking and more locally focused than EU spatial policy, it is appropriate to discuss it in the context of this chapter.

Crossborder planning, in the sense of crossborder co-operation and co-ordination of plans between local and regional planning authorities on either side of a national frontier, has a long history and is relatively routine. It is the first aspect of spatial planning to feel the effects of the original creation of the common market, and in urbanised border areas of the original six member-states, for example around Aachen, Saarbrücken and Strasbourg, it is a form of strategic city-region spatial planning and in some cases local planning that is a special case only because a national border happens to run across the city-region. However routine, this situation does of course create institutional and practical problems, from differences in planning systems and procedures to the fact that any telephone call across the agglomeration is charged as an international call.

As a result of European integration, many other cities which have not traditionally been involved in crossborder planning but which are located near the edge of their respective national territory have felt the need to develop crossborder planning strategy. This includes a number located on maritime borders, especially on short sea crossings which carry a great deal of traffic.

With the creation of the SEM and the growth in intra-EU trade and transport, development pressures on one side of a border can very easily spill over, generating pressures and influencing land and property markets on the other side, as enterprises take advantage of the four freedoms. Co-ordinated crossborder planning to anticipate and channel such pressures, therefore, is desirable.

Such pressures can also occur on external EU borders, especially on the border with former communist CEE countries, since economic and market conditions can vary dramatically and companies seek to balance the advantages of cheaper labour and/or land costs with the greater stability and access to markets of the EU member. Crossborder planning studies have been undertaken along the German–Polish border and the Vienna–Bratislava–Gyor region, for example. Finland has the longest border with a former communist country, and the only one with a TACIS country. Much of this length is with the Republic of Karelia, and the programme for this border area is outlined below.

The SEM is responsible for another form of border-related planning issue, that of economic development to compensate for the loss of employment among border control officials and support services. This is especially true of those member-states that have implemented the Schengen Agreement (see Chapter 2). Not only is official employment in the customs services much reduced, but the large number of traders who relied on the fact that travellers had to stop anyway, so took advantage of local services, have also lost much of their custom. Development programmes to provide an alternative economic base are likely in these circumstances to be formulated on a crossborder basis. One category of business that still is necessary on any border is that of currency exchange bureaux. Their trade will also disappear if and when a single currency is introduced.

Crossborder planning co-ordination is also necessary across a number of relatively sparcely populated border regions across which new transport links are to be constructed. Most member-states have a legally binding system of planning. In such a system, the line of the proposed route of a new road or railway must be specified in the appropriate form of a local or building plan on both sides of the border, and these plans approved before the proposed development is authorised and construction can take place. Since a major purpose of the TENs programme is to fill in the missing links in the EU's transport network, this situation will necessarily occur in the implementation process. The Compendium project (see Chapter 12) will assist in overcoming such problems by providing an information base.

Europe 2000+ points out that the EU of 12 had roughly 10,000 km of borders, 60% of which was intra-EU and 40% external. The latter it classifies into four groups: those with EEA countries, most of which became internal borders after the 1995 enlargement; those bordering Switzerland where some restrictions will continue as Switzerland is neither in the EEA nor a candidate for membership; those bordering CEE countries; and those Mediterranean countries which have comparatively narrow stretches of sea separating them from former communist or north African countries (Commission, 1994a).

Border regions adjacent to former communist countries face the greatest problems owing to an accumulation of factors: opening of a previously sealed border; sudden transition to free movement of goods and people; and very marked differences in terms of income, employment opportunities, environmental standards, property prices, planning systems, etc. Great economic disparities and migratory pressures are a feature of these regions and even more of those areas accessible from north Africa.

INTERREG II

INTERREG is one of the Community Initiatives funded by the Structural Funds (see Chapter 7). It was adopted in 1990 to assist in the preparation of border areas for the removal of internal frontiers under the SEM. In its first period it concentrated on land borders, funding 31 Operational Programmes. Of these, 24 were for internal border areas. Activities supported included most forms of economic and rural development, transport, trade and tourism, and environmental co-operation.

The Edinburgh Summit in December 1992 resolved that priority should be given to the type of activity included in INTERREG during the 1994–9 programming period, endorsing the importance attached to it by DG XVI in achieving their overall objectives. INTERREG II, for 1994–9, is the biggest Community Initiative with a budget of 2.4 becus of which 1.8 becus is for Objective-1 regions. This represents 21% of the whole Community Initiative budget.

INTERREG II has two distinct strands because it combines the functions of INTERREG I and REGEN. The INTERREG component aims to develop crossborder co-operation and to help areas on the EU's internal and external borders overcome specific problems arising from their position. Rather more maritime border areas are eligible than was the case with INTERREG I, and rather greater geographical flexibility is allowed in defining areas to be included. All NUTS-3 areas in the EU situated on internal and external land borders, plus certain NUTS-3 areas on sea borders, are eligible for INTERREG II. In some cases, areas adjacent to those qualifying may also be included in order to create a more rational spatial planning area.

It has increased flexibility in other ways as well, as the range of eligible activities has been extended to include education and health, media services, language training, spatial planning and measures supplementary to TENs (Commission, 1994a: 129). For external borders, the PHARE programme has a budget of 150 mecus for EU–PHAREland crossborder co-operation. This is mostly concentrated on crossborder transport infrastructure and environmental improvement projects. Another new element in INTERREG is funding to support and evaluate INTERREG-PHARE initiatives.

The REGEN element aims to fill gaps in natural gas and other energy networks and provide interconnections with wider European networks. The link between the Greek and Italian electricity networks noted in Chapter 7 is one of four specific schemes supported in the 1989–93 phase of REGEN, the others being a natural-gas pipeline between the UK and Ireland and gas distribution networks in Greece and Portugal (see Chapter 9).

Crossborder planning in practice

There are many examples of successful crossborder co-operation in planning, several of which predate the INTERREG programme. Those around the borders of Belgium and The Netherlands are among the most developed. Joint development plans have been adopted in a number of cases, some of which are even legally binding and therefore have the same legal status as plans within one country (Commission, 1994a: 134). Such plans are most helpful on borders crossed by major transport routes, for example around Aachen and the Menton–Ventimiglia area. Joint planning has also led to joint development of business projects such as the European Development Centre at Longwy and the Aachen–Heerlen business park.

Although the city of Maastricht enjoys a high degree of name recognition, it is not in fact a large city, having a population of around 120,000. In relation to Dutch national space it is highly peripheral, but in the EU it occupies a central location within an urban region of 3.5 million people in three countries. The other major cities of this urban region, referred to as MHAL, are Aachen in Germany, Liège (Wallonia) and Hasselt and Genk (Flanders) in Belgium, plus Heerlen in The Netherlands. In anticipation of the SEM and in response to the Dutch fourth report extra on national spatial policy (see Chapter 6), in 1989 the responsible ministers signed an agreement to co-operate on the development of spatial plans for the physical and economic development of the MHAL urban region, considered as one planning region.

An International Co-ordinating Commission was set up with the task of preparing a long-term vision for MHAL, in association with the secretariat of the Benelux Union. The main themes are

- urban development and co-operation
- tourism and recreation
- economic development
- traffic and transport
- quality of the environment
- the rural areas between the urban centres.

A high level of co-operation enables the MHAL region to promote itself as an integrated 'Euroregion' that is highly accessible, well served by international airports and the future high-speed rail system, and has a major concentration of European academic and research institutions and many international businesses. The presence of borders does not prevent the infrastructure of the region serving the whole MHAL area, whichever country it happens to be located in.

On a broader scale, the complexities of planning for the Benelux borders are greatly eased by the work undertaken for the Outline Structural Plan for the Benelux Union (see Chapter 6) and the Guideline Planning Concept for the German–Dutch border. This is commended in *Europe 2000+* for its bottom-up style and its concentration on those strategic issues for which crossborder planning is needed. By contrast, *Europe 2000+* is critical of the planning concept for the German–Polish border (one of the German government's initiatives to assist in the integration of CEE countries with the EU – see Chapter 6) for being the product of top-down planning from national and state authorities. This is an area of relatively recent co-operation where the mutual learning process still has some way to go (van der Boel, 1994).

One of the areas eligible for INTERREG for the first time under INTERREG II is west Wales and the eastern parts of Ireland (Figure 8.1). A learning process is also required but, in this case, the learning process does not concern planning as such but of identifying the ways in which to take fullest advantage of the opportunities offered by INTERREG.

In spite of their common Celtic heritage and tourist interests, many people find it difficult to think in crossborder terms as the funding criteria require. To stimulate ideas, the regional authorities in Ireland and the European Offices of Dyfed and Gwynedd County Councils in Wales are issuing newsletters and other publicity in both Welsh and English designed to draw attention to the new opportunities now available, in which the example of Kent–Nord-Pas de Calais–Belgium (see below) is quoted to show what can be done.

The major criteria for project approval are that projects must be crossborder and contribute to economic development while respecting the environment. Projects of maritime development and general economic development will be sought. These may involve such things as joint research into the marine and coastal environment, improvement in transport infrastructure and information systems to overcome capacity and peripherality problems, marketing services for SMEs, and joint tourist development packages which focus on the common Celtic links and encourage visitors to explore the rich cultural heritage on both sides of the Irish Sea.

Figure 8.1 *The Celtic Interreg, Ireland and Wales*

Euroregion Kent–Nord-Pas de Calais–Belgium

The authorities linked by the Channel Tunnel, the county of Kent in England and the *région* of Nord-Pas de Calais in France, have worked together for some years since the proposal to build the tunnel was given the go-ahead in 1986. More recently, they have formed 'Euroregion' and registered it as a European Economic Interest Group. It is therefore one of the more mature maritime border areas and has formed part of both INTERREG I and INTERREG II.

Euroregion consists of five local and regional authorities, Kent, Nord-Pas de Calais and the three regional governments of Belgium, Brussels-Capital, Flanders and Wallonia. These regions all have a common interest in the development of transport infrastructure and of development and environmental pressures that follow from that. Within this grouping are a number of localities that have suffered from peripherality, or which fear that the economic benefits of the new infrastructure may pass them by. They take the view that there are a number of opportunities which can be exploited by working together on economic development, transport, communications and environmental planning. A joint team has produced a strategic planning document called *A Vision for Euroregion* in which a number of issues for further study and joint policy development are set out. These include demographic change, economic opportunities, integrated development of road, rail, water and air transport, adherence to an agreed environmental charter, and spatial planning policies for urbanisation, historic town centres and limiting urban sprawl.

Oresund

Another urbanised region to be united by a fixed link is the Oresund region around Copenhagen and Malmö, identified in *Denmark 2018* (see Chapter 6) as one of the major urban centres. The concept is elaborated in *The Oresund Region – a Europole* (Ministry of the Environment, 1993). The Oresund region is defined to include Malmöhus County, including the city of Malmö, on the Swedish side, and the capital region, including Copenhagen, Frederiksberg and the Counties of Roskilde and Frederiksborg. This area roughly corresponds to the commuting areas around Malmö and Copenhagen.

North Karelia

The Finnish government intends to make full use of INTERREG, and the Regional Council of North Karelia has prepared a programme for the border area adjacent to the Republic of Karelia, an autonomous republic of the Russian Federation. Proposals for INTERREG are to be co-ordinated with TACIS funding in the Republic of Karelia on the Russian side of the border. In addition to INTERREG funding, this programme will make use of the LEADER, SME and URBAN Community Initiatives.

A set of six objectives and priorities have been set for north Karelia around the themes of expertise in forestry; specialised industrial expertise in the metal and plastics industries which are major local employers; tourism and leisure; diversification of rural industries; communications networks and energy; and co-operation with nearby parts of the Russian Federation. The region has one of the international crossing points at Niirala and promotion of its use is important to the local economies on both sides. An Operational Programme has been developed around these themes with 17 specific measures, some of which are specific to certain of the Community Initiatives that contribute to the north Karelia programme.

Joining in

Inter-regional co-operation presents many challenges – above all, to the imagination. It is easy for policy-makers to be set in their ways and inhibited in their thinking, not receptive to new mental maps and removal of cartesian inhibitions. Networking is a way in which this challenge is being met, and it plays a part in developing integrated policy frameworks for Europe as a whole. However, it is necessary for policy-makers to learn to 'think European'.

There is an EU programme under Article 10 to help new participants, the Exchange of Experience Programme. This is administered on behalf of the Commission by the Council of European Municipalities and Regions and the Assembly of European Regions, now called PACTE. The PACTE programme aims to promote small-scale projects of transnational co-operation, and gives priority to local and regional authorities wishing to participate for the first time. It is quite flexible, not specifying particular themes in order to encourage new ideas and participants who may later join in bigger projects. The main stipulation is that PACTE projects must normally involve a partner from an Objective-1 area. Even in starter programmes such as this, the overall spatial objective of cohesion is not forgotten.

9

Transport and trans-European networks

Transport policy, unlike regional or environment policy, was specified in the Treaty of Rome. The authors of the treaty recognised that transport, along with agriculture, would play a vital role in the creation of a common market. Furthermore, transport infrastructure is normally a key component of any strategic spatial policy, and certainly is at the EU scale. Why, therefore, is EU transport policy not one of the first components to be discussed in any overall review of spatial policy?

One reason is the earlier emphasis on technical harmonisation and safety legislation. Another is that there was less political momentum, in comparison with the early drive to establish a CAP and with the ECSC programme for the coal and steel sectors. This was not for want of pressure from the EP, or lack of interest on the part of the Commission. The Commission was conscious of the failure to implement the Treaty of Rome in respect of a common transport policy, and in 1979 issued a memorandum on the role of the Community in the development of transport infrastructure (Commission, 1979) which was supported by the Klinkenborg Report adopted by the EP (see Chapter 5). As Chapter 5 pointed out, lack of action by the Council of Ministers on transport policy led the EP to seek a legal ruling in the European Court of Justice. Following this, progress began in the context of the SEM programme and the picture looks set to change more dramatically in the 1990s.

The Maastricht Treaty introduced a new title: Title XII: Trans-European Networks. This requires the EU to promote the development of TENs in the areas of transport, telecommunications and energy infrastructures. At the Edinburgh Summit in December 1992, agreement was reached on the European Investment Fund and other financial instruments to promote a programme of investment in these forms of infrastructure, explicitly to fill missing links in the communications networks of the EU. Measures under this title may also be financed by the Cohesion Fund established under Article 130(d) of the treaty.

The portfolio for transport within the Commission which took office in January 1995 was given to the new UK Commissioner, Neil Kinnock. This explicitly includes responsibility for TENs. The new powers for TENs, together with the new commission, have generated a new momentum for transport policy which will have particular emphasis on the spatial dimension.

In fact, TENs are seen by some to have the potential to be the latest 'big idea' that might capture the public imagination in the way that the 1992 single-market programme did. In his 1993 white paper on *Growth, Competitiveness and Employment*, Jacques Delors, then President of the European Commission (see Chapter 5), devoted a whole chapter to TENs, identifying them as a means of promoting growth and creating new jobs through an investment partnership of public and private capital.

They are also seen as essential if the objective of ensuring economic and social cohesion of the EU is to be achieved, and they will also play a vital role in the programme to transform the economies of the CEE and reorientate them westwards. TENs are in fact a means by which the original objectives of the Treaty of Rome, and the European integration sought by its promoters, may be brought closer. Just as the Roman roads were essential to bind the Roman empire together, and the railways were essential to the continental expansion of the USA in the nineteenth century, so TENs will bind Europe together with a mixture of old and new transport modes and communication technologies.

In the context of spatial planning at the EU scale, TENs have an obvious resonance. If asked what spatial planning issues legitimately should be addressed in a spatial plan for the whole of the EU, it is likely that most local planners or other citizens would readily accept that transport infrastructure is one such issue. If this is so, the relationship between transport TENs and the development of EU spatial policy needs to be clear if the latter is to achieve general acceptance.

Context

The 1993 white paper supported TENs in the wider context of the economic development of the EU as a whole. Earlier green and white papers had also been issued in 1992 by DG VII as a basis for policy development. The green paper, *The Impact of Transport on the Environment. A Community Strategy for 'Sustainable Mobility'* (Commission, 1992c), was jointly prepared by the transport DG and DG XI (Environment) and was also a follow-up to the green book on the urban environment (Commission, 1990b; see Chapter 10). Later in 1992, the white paper entitled *Commission Communication on the Future Development of the Common Transport Policy* was issued (Commission, 1992d).

The white paper is quite clear about the proper relationship between transport policy and spatial policy, noting that in Objective-1 areas the capacity of transport infrastructure is only in the range of 50–60% of the EU average: 'There is a close interaction between the development of transport and the spatial distribution of economic activity' (Commission, 1992d; 86). It goes on to state that 'the substantive goal of Community action is defined as the establishment and development of a trans-European transport network, within a framework of a system of open and competitive markets, through the promotion of interconnection and interoperability of national networks and access thereto' (p. 152).

The strategy outlined in the white paper contains certain key elements:

- Strengthening the single market in respect of free movement of goods and passengers.
- Developing genuinely integrated transport systems with particular emphasis on combining different modes and better interconnections between modes.
- Improving infrastructure, links between peripheral and central parts of the EU and developing TENs.
- Using charges and fiscal incentives to promote use of more environmentally friendly modes, and stricter pollution control.
- Improving safety, especially on roads.
- Social measures to ensure better working conditions and training for technological changes for employees and better access for the disabled.
- Development of the external dimension of transport policy in relation to third countries, especially within the EEA and CEE.

TENs form the most visible and dramatic part of the strategy, and are discussed at length below. The external dimension of transport policy, and especially its role in contributing to the transformation process of the CEE countries and their future integration into the EU, is a part of the TENs strategy. First, some of the other aspects are considered.

SEM measures

Seventeen transport measures have been adopted under the single-market programme (Roberts *et al.*, 1993), several of which were concerned with air and maritime transport, and several with road transport. The main emphasis is on freedom of competition and liberalisation of service provision in any member-state, whether the operator is based in that state or not. The white paper indicates areas where state aids still exist which may distort competition and issues not yet tackled by the single-market programme such as liberalisation of taxi services, security transport and rented cars.

Air transport

Although the emphasis in this chapter is on surface modes of transport, air transport must not be forgotten in this context. Liberalisation of European air services has been an aspiration for a long time, with some governments (including the Dutch and UK) pressing hard for it against resistance from other governments and state airlines. In 1974, a ruling of the European Court of Justice that the airline industry was not exempt from the competition provisions of the Treaty of Rome opened the way for the protected-fare structure of the scheduled airlines to be challenged, and in 1984 a bilateral agreement between the Dutch and UK governments opened up their airspace to new carriers, leading very rapidly to fare reductions and a 10% increase in passengers (Owen and Dynes, 1993).

Integrated transport

The problem of inefficiency caused by imbalance in the provision of different forms of transport is highlighted in the white paper. Freight movements within the EU have increased by 50% over 20 years, 70% of which now goes by road as inland waterway and rail have lost market share. Passenger movement has increased by over 85% in the same period, and 79% of all passenger transport is now by private car. The white paper outlines a policy of combining the different modes of transport (road, rail, inland waterway, air and sea) in order to make more efficient use of infrastructure, allow greater choice and flexibility to the users and less disadvantage to the peripheral areas.

In order to tackle overcapacity in certain modes of transport and undercapacity in others, the white paper suggests the development of an EU framework for charging the users full costs for the use of the infrastructure, and for harmonising excise duties on fuels. In the absence of such charges, distortions in the perceived cost to the user occur not only between modes such as road and rail but also between different national networks. These distortions can have spatial implications at the EU scale. For example, the imposition or lack of motorway tolls in different member-states may affect the routeing decisions of road haulage operators as well as holiday motorists.

Environment

The need for environmental considerations to be integrated within transport policy as part of the global strategy set out in the *Fifth Action Programme on the Environment* (Commission, 1992b; see Chapter 10) is recognised. It is not sufficient to reduce operational pollution such as exhaust emissions and noise. The global strategy must address demand for different modes of transport, promote efficient use of existing networks and of environmentally friendly forms of transport, and include spatial planning principles designed to minimise demand and dependency on private vehicles. These issues are discussed further in Chapter 10.

Title XII: Trans-European networks

This section of the treaty consists of three articles, Articles 129(b), (c) and (d), and is of potentially great significance. The objective is to enable the EU to promote and contribute financially to the establishment and development of networks for transport, telecommunications and energy infrastructure to enable 'economic operators and regional and local communities to derive full benefit from the setting up of an area without internal frontiers' (Article 129(b1)). 'Action by the Community shall aim at promoting the interconnection and interoperability of national networks as well as access to such networks' (Article 129(b2)). The Commission sees this section as playing an important role in the development of an integrated European economy and transport structure, extending to islands, landlocked and peripheral regions. The spatial planning process accompanying the development of proposals for TENs will require the capacity to think innovatively and conceptually at the European spatial scale.

Title XII gives the Commission competence to propose measures designed to achieve these objectives through integrated operation of national transport, telecommunications and energy infrastructure networks. Special account of islands, landlocked and peripheral regions of the Community are to be taken into account in such measures. Transport infrastructure development relates to five modes of transport: rail, road, inland waterways, sea and air, plus infrastructure based on integrated combinations of these.

The treaty also provides that the Community may co-operate with third countries to undertake projects in order to achieve the objectives of TENs. This clearly will allow for integration of the EFTA countries in such networks, for example by new transport links, and measures to assist the new German *Länder* may also benefit some of the former Communist Party states. The main purpose of extending TENs to third countries is to support the wider and longer-term integration of the whole of Europe and to support the economies of poorer neighbours in the CIS and in north Africa.

Even within the heartland of the EU in countries such as Belgium and The Netherlands, the extent to which rail networks, for example, are national rather than European is quite noticeable. The provision for TENs therefore provides a treaty basis for policy development along the lines outlined in the *Europe 2000* report (Commission, 1991) for high-speed rail and major road networks. Proposals for road, rail and inland waterways, and guidelines for these and other sectors including airports and seaports have been made by the Commission (Commission, 1992e; 1993d), as have proposals for rules and criteria governing Community financial aid for TENs (Commission, 1994d).

TENs

The objective of TENs is to provide comprehensive networks of physical and communication infrastructure for the whole of the EU. The key features of these networks are that they will be interoperable, they will interconnect, and they will be European not national in their spatial form.

TENs are seen as a powerful force for informal integration. They are intended to facilitate the Europeanisation of national economies, to improve mobility by promoting the interaction or interdependence of existing member-state networks and, in a sense, thereby to create 'a Europe of networks'.

The scope of the TENs programme goes beyond conventional transport to include energy and telecommunication networks. Originally, it was proposed to include vocational training networks to facilitate open and distance learning, but this was later deleted. Modes of transport to be included in TENs include road, rail, inland waterway, air and intra-EU shipping.

The emphasis on the peripheral and island parts of the EU and the role of TENs in economic and social cohesion is based on theory that infrastructure development stimulates economic growth and that telecommunications can overcome the distance effect by offering a substitute for travel, so removing some of the disadvantages of peripheral and island locations.

The TENs programme is, at the time of writing, still in preparation. Three sets of action are being developed:

1. Guidelines concerning objectives, priorities, identification of projects of common interest between member-states.
2. Measures to ensure interoperability of networks.
3. Financial support.

Finance

TENs will require the investment of very large sums of money. The 1993 white paper (Commission, 1993d) estimated that total investment required by 1999 to implement the programme would be up to 400 billion ecus, 220 billion on transport, 150 billion on telecommunications and 13 billion on energy. It went on to calculate that member-state budgets (including resources available to national, regional and local authorities) might be able to offer up to one-third of this but that there was no way that the public purse could bear the total cost. The EU budget line itself is intended to represent only 1% of the total cost of the TENs programme, and a ceiling has been agreed of 10% of the cost of any one project that can be funded by the EU for projects of common interest. Unlike the Structural Funds and the Cohesion Fund, therefore, the EU's TENs budget will not make anything happen that could not otherwise go ahead on the basis of national sources of public or private finance. Financing TENs will take the form of a public-private partnership within and between member-states, with the Commission taking the role of facilitator and co-ordinator.

For any project, the sum allocated from the EU budget must be matched by public funds from national or local sources. Any project that is capable of going ahead as a purely commercial proposition with private finance is not eligible for funding from the TENs budget. In general, it will be telecommunications projects that are most likely to be privately financed as commercial projects.

Directly funded EU action under the TENs budget will be confined to certain feasibility studies and demonstration projects. Up to 50% of feasibility studies may be funded by the Commission so long as this does not exceed 10% of the total project cost.

There are other EU funding sources which will be available for TENs projects, however (see Chapter 7). For projects in the four Cohesion Fund countries (Greece, Ireland, Portugal, Spain), up to 85% of capital costs may be available. Following agreement at the Edinburgh Summit of 1992, Finance may also be available from the European Investment Fund and the EIB through the so-called Edinburgh Facility (McDonald, 1995).

Start-up phase

In order to start the process of stimulating public-private partnership and demonstrating the scale and range of what may be proposed under the TENs programme, the Commission argued that a set of priority projects should be identified. The available budget would be concentrated on these in the first phase in order to demonstrate the concept and encourage private investment.

The Copenhagen Summit in June 1993 appointed two working parties to identify priority projects. Martin Bangemann, as then Commissioner for Information Technology, reported on the telecommunications sector, and Henning Christopherson, then Economic and Financial Affairs Commissioner, chaired a group of national government representatives to consider transport and energy.

The Christopherson Report

The Christopherson working group produced a proposed short-list of projects from the longer list of 26 transport projects, 37 electricity and 29 gas projects suggested in the 1993 white paper (Commission, 1993d). The short-list, which was presented to the Corfu Summit in June 1994, consisted of 11 transport projects in a relatively advanced stage of planning plus five links outside the EU to CEE countries, and seven internal energy links plus one outside the EU. The 11 priority projects, plus 10 which could be accelerated to start in two years and 12 which require supplementary studies, are listed in *Europe 2000+* (Commission, 1994a: 56–7). At the Corfu meeting, the transport links were in principle accepted and member-states were asked to press on with preparation and planning procedures so that, where possible, they could be ready for implementation to start in 1995 or 1996 at the latest.

Christopherson was asked to review the external transport and energy proposals for final approval at the Essen Summit in December 1994. Meanwhile, extensive lobbying by national and subnational governments and transport interests took place. The Essen Summit devoted much of its attention to future eastward enlargement of the EU, and the heads of the governments of the Visegrad countries participated. Consideration of the TENs projects therefore took place in this context. The outcome was a list of 14 transport projects (listed in Figure 9.1) and eight energy projects within the EU (listed in Figure 9.2). Beyond the EU, two external energy projects were approved and a number of proposals were agreed for further detailed study.

Transport projects

The 14 priority transport projects show a distinct bias towards rail: some are combined road/rail, three are road only and there is one airport improvement but no inland waterway project (see Figure 9.1).

High-speed train/combined transport: Verona–Berlin

This project aims to meet a number of objectives by developing a high-speed rail route from northern Italy via the Brenner Pass in Austria to München, Nürnberg, Erfurt and Halle/Leipzig to Berlin. The volume of goods transported from northern Italy to Germany has increased greatly, and the ecologically sensitive Austrian Alps have borne the heaviest pressure. The growth in demand has been particularly great in the 1990s with the completion of the SEM programme and the addition of traffic from Greece seeking to avoid the Yugoslav conflict by crossing to Brindisi and transitting through Italy.

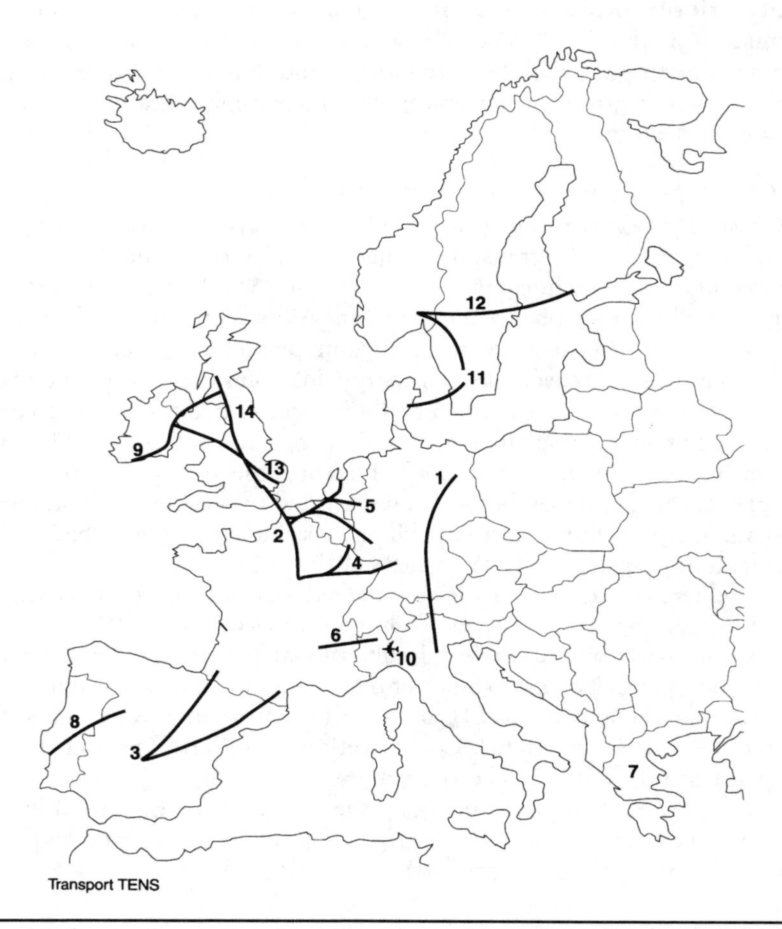

1. High-speed train/combined transport: Verona–Berlin
2. High-speed train: Paris–Brussels–Cologne–Amsterdam–London
3. High-speed train south
4. High-speed train east
5. Conventional rail/combined transport: Rotterdam–German border
6. High-speed train/combined transport: Lyon–Turin
7. Motorways: Greece
8. Motorway: Lisbon–Valladolid
9. Conventional rail: Cork–Dublin–Belfast–Larne–Stranraer
10. Airport: Milan Malpensa
11. Fixed rail/road link: Denmark–Sweden
12. Road-rail/combined transport: the Nordic triangle
13. Road: Ireland–UK–Benelux
14. Rail: UK west-coast main line

Figure 9.1 *Transport TENs Priority Projects*

The Verona–München part of the project is for a new high-speed rail track, including a proposal for a 55-km rail tunnel through the Alps to relieve pressure on the Brenner Pass and increase capacity on the route. Prior to accession, the Austrian authorities tried to place limits on the number of lorries using the road routes across the Tirolean Alps and divert them on to a piggy-back service on existing rail routes, which was rather slow and unpopular.

The link to Berlin involves a mixture of new track and track upgraded to high-speed standards, as part of the strategy to improve communications in the new *Länder*, make Berlin more accessible and open the way to eventual extension into the CEE.

High-speed train: Paris–Brussels–Cologne–Amsterdam–London

The aim of this project, or set of projects radiating from Brussels, is to make rail travel competitive with air travel on the very busy routes between these cities, and free existing conventional track to take a higher volume of freight.

High-speed train south: Madrid–Barcelona–Perpignan–Montpellier and Madrid–Vitoria–Dax

This project consists of two high-speed rail routes linking Madrid via Valladolid and Vitoria to the French TGV Atlantique at Dax, and via Barcelona to the Mediterranean route at Perpignon. These will be at the European standard gauge, eliminating the need for carriage changes at the border. Within Spain, four hours is expected to be cut from the Madrid–Barcelona journey time.

High-speed train east: Paris–Metz–Strasbourg–Appenweiler; Metz–Saarbrücken-Mannheim; Metz–Luxembourg

This is a project already agreed between the French and German governments which will improve communication in the core of the EU by joining up national networks. New high-speed track is to be built from Paris via Metz to connect with the German high-speed network in two places, near Strasbourg and near Saarbrücken. This project is an element in a wider strategy: a spur will be built from Metz to Luxembourg which could later be extended to Brussels; and on the German side, the route is planned with eventual extension to Frankfurt and Berlin in mind.

Conventional rail/combined transport: Rotterdam–German border

The actual project involves only 160 km of new track from Rotterdam to the German border, with the wider objective of connecting with existing transport (including the Rhine ports) in order to increase the capacity to transport freight from Rotterdam to the industrial conurbations of the Ruhr, Main and Neckar agglomerations.

High-speed train/combined transport: Lyon–Turin

This is also a major example of completing a missing link which will have spatial implications well beyond its actual location. The project is for a 250-km route of new and upgraded rail track linking the prosperous northern regions of Italy with the French north–south TGV. By doing so, it will eventually connect Italy with Spain via Dax and Montpelier (project 3) and with the central group of capital cities (project 2). Eastwards, it will link with the north–south route from Brindisi, the main port for Greece, with connections to Slovenia and Austria.

Motorways: Greece
Although entirely within Greece, this project is itself a network with connections from the Turkish and Bulgarian borders to the western port of Igoumenitsa and among Patras, Athens and Thessaloniki.

Motorway: Lisbon–Valladolid
By upgrading and realigning an existing road to motorway standard, an alternative to existing heavily congested routes via Madrid will be provided.

Conventional rail: Cork–Dublin–Belfast–Larne–Stranraer
As the only crossborder rail link between Ireland and Northern Ireland, this project is seen as a dividend of the peace process. It will involve upgrading to increase speed and capacity, although not to speeds which would justify designation as a high-speed track.

In addition to the Larne–Stranraer ferry routes, this is intended to improve access to the Dublin–Holyhead ferry route and so to continental Europe via the Channel Tunnel. In the Christopherson Report, it was characterised as the cross on a capital T, the stem being the Holyhead–Channel Tunnel route. A connection with projects 13 and 14 is also part of the rationale, again in order to improve Irish access to the central markets of the EU.

Airport: Milan Malpensa
Malpensa is the second airport of Milan, located 50 km away from the city. It is to be upgraded to form an international hub for northern Italy, concentrating on long haul and leaving Milan's other main airport, Linate, for short-haul flight, most of which will be internal to the EU. A project involving only a single airport may hardly fit the definition of a network, but it was included in the priority programme partly because otherwise the airport sector would not have been represented.

Fixed rail/road link: Denmark–Sweden
This project is one of those which will affect the spatial structure of Europe as a whole (see Chapter 6). It consists of a 15.5-km length of four-lane motorway and double-track railway from Copenhagen to Malmö across the Oresund. A 3.8-km tunnel will take the route from the Danish coast to an artificial island which will serve as the western support for a 7.5-km bridge to Sweden, 1.1 km of which will be elevated to allow shipping to pass. Work began on the access links on the Danish side in 1993, although final approval of the project by the Swedish Parliament was delayed because of environmental objections and had not at that time been given.

Road-rail/combined transport: the Nordic triangle
This is a network of road and rail links among Oslo, Stockholm and Copenhagen (via the Oresund link), and from Turku to Helsinki in Finland. Consideration is being given in Finland to the possible extension to the Russian border. The Norwegian sections will need to be the subject of specific co-operation agreements between Norway and the EU.

Road: Ireland–UK–Benelux
This project seeks to connect the Cork–Dublin–Belfast road and rail routes via the ferries to Holyhead and Stranraer and the British motorway system to the port facilities at Felixstowe and Harwich to create a through route to Rotterdam and Zeebrugge. Some relatively short sections of new road construction is involved but much of the project involves realignment and road widening.

Rail: UK west-coast main line
The existing intercity route from London to Glasgow, also serving Coventry, Birmingham, Liverpool, Preston and Carlisle, is the basis of this project. It is to be improved to allow for high-speed operation at up to 225 kph with improvements to the track and electrification and resignalling. As with the previous project, part of the rationale for inclusion in the list is to overcome Irish peripherality. Piggy-back freight transport services will be developed.

Energy projects

The eight priority projects are listed in Figure 9.2.

Electrical interconnection: Greece–Italy
Greece is a net importer of power with interconnections only over its land borders. Limited surpluses and a lack of recent investment in new generating capacity have left Greece susceptible to blackouts, and a link to Italy will provide security against this. The project itself is a submarine cable from Galatina in Italy to Arachtos on the Greek mainland, plus connecting transmission lines. In

1. Electrical interconnection: Greece–Italy
2. Electrical interconnection: France–Italy
3. Electrical interconnection: France–Spain
4. Electrical interconnection: Spain–Portugal
5. Electrical interconnection: Denmark
6. Natural-gas network: Greece
7. Natural-gas network: Portugal
8. Natural-gas network: Portugal–Spain

Figure 9.2 *Priority Projects for Energy TENs*

addition to the TENs programme, funding will come from the ERDF and the REGEN programme (see Chapter 7).

Electrical interconnection: Italy–France–Spain–Portugal; Denmark
The electrical interconnection between France and Italy links the Haute Savoie with Piossasco, west of Turin, enabling France to export its surplus nuclear energy and respond to the Italian need to import electricity. The France–Spain and Spain–Portugal links are also to allow trading and balancing of supply and demand, but without such a clear-cut flow. In the case of Denmark, the mainland is part of the continental network, predominantly based on thermal generation, while the islands form part of the Scandinavian Nordel network, which has an abundance of hydropower. By completing a cable link within Denmark, these can be linked – a project considered to be of wide strategic importance.

Natural-gas networks
In the case of Greece, the project involves both international links with supplies from Russia and Algeria, and a distribution network. Funding through the Structural Funds, CSF and REGEN is also available. The project in Portugal involves a pipeline from Setubal south of Lisbon to Braga in the north, bringing 85% of industry and 80% of the population within reach of a natural-gas supply for the first time. The Portugal–Spain link is a connected project which will allow natural-gas supplies to be introduced in Extremadura and Galicia. Both are extensions of one of the two third-country projects approved at the Essen Summit.

Third-country projects
The two energy TENs with neighbouring countries approved at the Essen Summit were both natural-gas pipelines, one linking Algeria and Morocco with the EU via the Straits of Gibraltar and the other linking Russia via Belarus and Poland to the EU.

The former will connect with projects 7 and 8 to provide Portugal with its first natural-gas supply and double the supply available in Spain. In a later phase, a pipeline will be laid to connect with the French natural-gas system. The latter will provide an alternative route to the existing pipeline from the Siberian gasfields via Ukraine, Slovakia and the Czech Republic to the EU in order to offer additional security and increase capacity. The first phase will involve a system of pipelines in eastern Germany and Poland, extending in the second phase through Belarus to Russia to connect with the existing link to Siberia. Russia is planning a substantial increase in production of natural gas on the Yamal peninsula in north-west Siberia, a quarter of which is intended to be supplied to the EU.

CEE

Transport is one of the key aspects of the programmes aimed at the transformation of the CEE countries (PHAREland and TACISland). Apart from the

1. Tallinn–Riga–Kaunas–Warsaw with a branch Riga–Kaliningrad–Gdansk
2. Berlin–Warsaw–Minsk–Moscow
3. Berlin–Wroclaw–Katowice–Lvov–Kiev and Dresden–Wroclaw
4. Dresden–Prague–Bratislava–Gyor–Budapest–Arad–Craiova–Sofia–Thessaloniki and Sofia–Plovdiv–Istanbul with a branch Nürnberg–Prague–Vienna–Gyor–Arad–Bucharest–Constanta
5. Trieste–Koper–Postojna–Ljubliana–Budapest–Uzgorod–Lvov, with branches Bratislava–Zilina–Kosice–Uzgorod and Rijeka–Postojna
6. Gdansk–Katowice–Zilina with a branch Torun–Poznan
7. The River Danube, including all ports in CEE countries
8. Durres–Tirana–Skopje–Sofia–Plovdiv–Burgas–Varna
9. Plovdiv–Bucharest–Kichinev–Ljubasivka–Kiev–Vitebsk–Pskov–St Petersburg–Helsinki, with branches Odessa–Ljubasivka, Kiev–Minsk–Vilnius–Kaunas–Klaipeda, Kaunas–Kaliningrad and Kiev–Moscow

Figure 9.3 *CEE priority transport corridors*

EU's programmes, those of the UNECE and the CoE's European Conference of Ministers of Transport (see Chapter 2) play an important role. Ministers agreed at their second pan-European conference in Crete in March 1994 that a Europe-wide transport policy should be based on the following elements:

- A social-market approach, with competition within increasingly open national markets.
- A multimode approach with alternatives to road transport developed in order to overcome problems of congestion, safety and environmental damage.
- Convergence of regulatory policies, simplification of border-crossing procedures and the establishment of the highest possible safety, energy and environmental requirements (Commission, 1995c).

As a first step, the conference endorsed a set of indicative guidelines laying down the main infrastructure corridors for road and rail transport in the CEE. On the basis of criteria such as economic viability and the potential to interconnect with the EU's TENs programme, to link EU regions with regions of third countries, and to facilitate international flows, a set of nine priority transport corridors have been identified, all of which have both road and rail components (Figure 9.3).

Link with spatial policy

The link between spatial planning for TENs and overall EU spatial policy is clearest in the case of the four cohesion countries. In the case of all TENs infrastructure proposed in these countries, EU finance comes from the

Cohesion Fund which is the responsibility of DG XVI. The regional policy directorate therefore has been able to ensure that the TENs strategy is consistent with the overall regional policy strategy. Conversely, proposals for transport infrastructure funding by the EU within the cohesion countries are acceptable only if they form part of the TENs strategy.

The strategies that have been developed for TENs infrastructure elsewhere in the EU have proceeded without the broader perspective of EU spatial policy because of the political urgency attached to the programme. DG VII was therefore responsible for establishing the European interest which would justify different proposals.

A section of the *Europe 2000+* report (Commission, 1994a) is devoted to the effects of TENs on regional development and to the link with spatial planning, but it is not based on the final list of 14 projects approved at the Essen Summit, as the text was finalised beforehand. It does, however, note a number of points which will need to be taken into consideration in formulating spatial strategies at both EU and national levels. Analyses of accessibility suggest the central parts of the EU will have their advantages reinforced while peripheral parts will remain dependent on a more limited choice of transport and communication methods and less well developed connections. Several peripheral parts of the EU will benefit from substantially improved accessibility and journey times, but these effects will be selective rather than global.

Apart from the Malpensa project, the priority programme does not involve air transport projects, although deregulation of air transport and encouragement of greater use of regional airports, plus possibly the development of new airport facilities in peripheral locations, could play an important role.

Another problem noted in *Europe 2000+* follows from concentration on major networks and the danger of overlooking the need for secondary networks at the local and regional scales. Many areas remain poorly connected to access points on the major networks, such as motorway intersections and high-speed railway stations. This problem does not necessarily arise only in locations peripheral to the EU. Amiens in northern France, for example, is very concerned at the possible consequences of being bypassed by the London–Lille–Paris–Brussels TGV routes, and part of the problems of the Objective-1 regions of Hainault in southern Belgium and Flevoland in The Netherlands is due to their lack of connection to the major networks, in spite of their central EU location.

For peripheral regions, the problems of isolation can be considerable. The report notes studies that indicate that large inland areas of Portugal, Andalusia, west Wales, Ireland and northern Scotland are more than two hours' travel time from a major network, as are many of the island communities in Greece.

An associated issue is that the environmental costs of TENs fall along the whole length of the physical infrastructure, whereas the economic benefits occur at the ends or the access points. TENs are therefore by definition spatially redistributive.

The question of secondary networks and their connection with TENs and major networks is therefore a key issue for spatial planning. Political imperatives have meant that the priority TENs programme could not wait for the completion of the *Europe 2000+* work, and had to be negotiated and approved concurrently. The need remains for more fundamental examination of the implications of different modes of transport for spatial policy at the EU scale since this is just as important an issue as it would be in the context of local and regional planning.

Planning and implementation

Having adopted the list of 14 pilot transport TENs in December 1994, practical problems of financing, planning and implementation are being faced in a number of cases, especially where the physical development has to be co-ordinated across national borders, and where there is limited experience of managing projects and where a substantial proportion of the funding comes from private sector participation.

The Commission notes two particular lessons that have emerged from the early experience of implementing the TENs programme, both of which have implications for the development of spatial planning expertise and knowledge of systems at the EU scale (Commission, 1995). One is that national administrations need to develop flexible approaches to the design, planning and development of projects in order to make the projects attractive to private investors and forge successful public-private projects. In some countries, the public authorities responsible for implementation are familiar with the degree of autonomy that sometimes prevails when a construction project is wholly funded from their own public funds, but are not familiar with the task of reconciling different interests and assembling a financial package of the sort that will be necessary for most projects. This is not just a question of financial management: a project must be seen to command public support and meet an identifiable demand, and the route approved through a publically accountable planning process, in order to command investor confidence.

The second lesson is that the absence of a common or compatible juridical structure handicaps the financing, planning and co-ordination of transfrontier development projects. Compatibility does not necessarily require harmonisation of planning systems (see Chapter 12). The Compendium project should provide sufficient information to enable the implementing authorities to overcome this problem, and if EU legislation were to be needed, it would provide the information on which a proposal could be based.

Telecommunications

The Bangemann Report, *Europe and the Global Information Society* (1994), identified the following applications of networking: teleworking; distance learning; telematic services for SMEs; road traffic management; air traffic control; electronic tendering; home connections for electronic newspapers and bulletin boards; and information exchange networks among scientific research centres and universities, health-care and other public authorities.

One conclusion of the Bangemann Report was that TENs in the form of telecommunication projects had sufficient potential for profit that financing and implementation could be left to the private sector. The role of the Commission would then be as facilitator and co-ordinator, and it would draw up proposals for any necessary deregulation.

Unresolved issues

It is too early to be able to form any conclusions about how the implementation of TENs will work out and whether they will achieve their economic development and cohesion aspirations, but it is clear that a number of spatial planning issues are raised by the programme.

The emphasis on partnership funding and lack of a substantial EU contribution outside the Cohesion Fund countries will create pressure to concentrate on routes that serve a well established existing market or transport flow because these are where it will be easier to attract private sector finance. Implementation of proposals of this sort could then proceed in advance of more imaginative proposals for routes that are designed to stimulate traffic growth along new corridors of movement, link regional centres in different member-states directly, or serve the many examples of peripheral or economically backward regions that exist outside the four Cohesion Fund countries.

Where implementation of TENs goes ahead within cohesion countries, this could add to the pressure on existing networks and infrastructure in neighbouring non-cohesion countries. Capacity constraints may become more severe if new infrastructure in the cohesion country succeeds in generating more traffic, and this could lead to political pressure on the authorities in the neighbouring country to improve their parts of the network.

The Commission is of course thinking in terms of the European scale of spatial planning when it considers the priority projects, and the routes it proposes are generally designed to link the major cities and regional centres. The regional and local planning authorities in the member-states must be aware of the need for secondary networks, which may be crossborder but often will not, in order to ensure that the smaller towns and rural communities have adequate infrastructure linking them to the access points on the TENs. Otherwise, there is the danger that the regional cities with good outside communications may prosper, allowing the disparities between them and the smaller places in their traditional hinterland to grow. This is largely a matter for the authorities in the member-states under the subsidiarity principle, but Article 129(b), quoted above, does give the EU a duty in respect of access to networks.

The nature of much of the transport infrastructure proposed is such that it would increase speed or capacity between set access points. These may be high-speed rail interchanges, new or improved ports and regional airports or even motorway interchanges. The problem of secondary networks is rather less with the latter, but all modes have this characteristic that the benefits concentrate around nodes rather than be equally distributed along the line of the route. High-speed rail has some characteristics in common with air transport in this respect, and is in fact very competitive with airlines for passenger transportation over distances of up to 800 km.

The environmental impact of new transport infrastructure, in contrast, does occur along the entire length. This will undoubtedly affect the political acceptability of some proposals. A procedure for strategic environmental assessment has been proposed by the Commission to respond to this issue (see Chapter 10), but this poses considerable technical and procedural problems and no agreement on an acceptable form has been reached in the Council of Ministers.

TENs are likely in some instances to test the levels of crossborder co-operation in planning and co-ordination of authorisation procedures. The *Compendium of Spatial Planning Systems* (see Chapter 13) will provide information which should help overcome this problem, and it is possibly with issues such as this in mind that every national report for the compendium is required to include case studies of both TENs and crossborder co-operation projects.

The list of priority TENs schemes includes several road projects although they require a disproportionate amount of land compared with other modes (see Chapter 10) and raise other environmental concerns, especially in respect of air pollution. It is quite possible that the emphasis will shift towards favouring more environmentally friendly modes, which are of course the modes for which concentration of economic benefits around the access points is also a characteristic.

TENs may in many parts of the EU be a necessary precondition for economic growth, but they will not always be sufficient by themselves to achieve this aim. Imaginative thinking on the part of spatial policy-makers, perhaps based on some of the innovative links and associations that are growing out of the networking and inter-regional co-operation programmes discussed in Chapter 8, is certainly required in order to realise the potential benefits.

10

Environment policy

There is now a substantial body of EU environment policy, by no means all of which needs to be considered here in the context of EU spatial policy. The aim of this chapter is to set out the principles underlying EU environment policy, to outline its scope and to show how this policy sector interacts with spatial policy-making both at the EU scale and at the level of local authorities responsible for planning.

In a sense, environment policy lies at the core of planners' interests, and was the first sector of EU policy to impact directly on the central land-use planning function of planning authorities. On the other hand, much of the body of environment policy is somewhat peripheral to the central theme of EU spatial policy and its development in the 1990s. This chapter seeks to explain this paradox by considering the place of land use within the policy and contrasting environment policy's centrality within the Treaty on European Union with the rather diffuse response to the 1990 *Green Book on the Urban Environment* (Commission, 1990b) and the lack of drive in environment policy since its publication.

For local planners responsible for the process of authorisation of development projects, the single most significant piece of EU legislation is the 1985 directive on the environmental assessment of projects (Directive EEC/85/337; see Haigh, 1989: 349–56). This is discussed below in order to set out its basic principles and show how it fits within the wider logic of EU environment policy. For local and regional authorities concerned with more strategic policies to achieve the objectives of sustainability and balanced regional development, EU environment policy provides the frame of reference within which other programmes must fit.

Policy-making by DG XI, Environment, Nuclear Safety and Civil Protection, has traditionally been heavily orientated towards a top-down legislative approach, with local and regional authorities not much involved until directives were transposed into national legislation and they were called upon to implement them. Around 300 pieces of EU environment legislation had been approved by the mid-1990s. The domination of this approach had led DG XI to be sometimes characterised as a legislation factory.

There is some evidence that a change from this style is taking place during the 1990s, with a greater emphasis on partnerships and on basing policy initiatives on networking and links with local and regional authorities. Where legislation is envisaged, it is recognised that, as the 1992 *Green Paper on the Impact of Transport on the Environment* notes, the legislative role may fall to member-states or regional governments (Commission, 1992c: 5). This is partly a response to calls to demonstrate subsidiarity, and partly a reaction to the experience of the *Green Book on the Urban Environment* initiative (Commission, 1990b). The place of this important initiative in environment policy is explained below, but see Chapter 11 for a discussion of its substantive content.

Having been the fastest-growing sector of EU policy and legislation in the 1980s, the scope of EU environment policy is enormous. The momentum was at its greatest around the time of the 1989 European Year of the Environment and EP elections (see Chapter 5). Much of this legislation is concerned with specific pollutants or sectors of the environment, and it is possible here to do no more than provide signposts and guidance through its complexities and indicate sources of information, reference and further reading. Detailed discussion is confined to measures that form part of the spatial planning system in a direct sense, such as environmental assessment.

The first part of the chapter is concerned with principles and the place of environment policy in the treaties, the second with specific programmes and initiatives, and the third with issues and prospects.

The place of environment policy in the EU treaties

As Chapter 5 showed, environment policy dates back to the Paris Summit of 1972 but its inclusion in the treaties dates back only to the Single European Act of 1987 when Title VII on the environment was added. This was enhanced in the environment title, now Title XVI, in the Treaty on European Union (TEU).

The TEU is quite clear about the place that environmental considerations must have in all EU policy instruments and EU legislation. Title XVI, the environment title, sets general objectives to which EU policy on the environment *must* contribute:

- preserving, protecting and improving the quality of the environment;
- protecting human health;
- prudent and rational utilization of natural resources;
- promoting measures at international level to deal with regional or world-wide environmental problems.

(EC Council, 1992: Article 130(r), para. 1)

The environment title goes on in the second paragraph to state that, *inter alia*,

> Community policy on the environment shall aim at a high level of protection . . . shall be based on the precautionary principle and on the principles that preventive action should be taken, that environmental damage should as a priority be rectified at source and that the polluter should pay. Environmental protection requirements *must* be integrated into the definition and implementation of other Community policies.
>
> (EC Council, 1992: Article 130(r), para. 2, emphasis added)

The third paragraph of Article 130(r) requires that environment policy should be prepared to take into account

- . . . available scientific and technical data;
- environmental conditions in the various regions of the Community;
- the potential benefits and costs of action or lack of action;
- the economic and social development of the Community as a whole and the balanced development of its regions.

(EC Council, 1992: Article 130(r), para. 3)

All member-states can therefore be said to have accepted, when they ratified to treaty, the obligation laid down in the treaty to give high priority to the environment, to seek high environmental standards and to accept the incorporation of these in all other EU policies. As Louis-Paul Suetens, founding chair of the European Environment Bureau and its Honorary President, states: 'It is tempting to interpret and construe these broad terms so as to proclaim the protection of the environment as an objective which completely overrides all other objectives. But this view needs nuancing' (EEB, 1994: 24).

This caution is based on the first bulleted point of para. 3 above, which allows some scope for flexible interpretation by different member-states, and on the second, referring to the costs and benefits of action. Nevertheless, environment policy is now firmly established as a major policy sector, one whose profile may recede or be more prominent from time to time but which will not go away.

The connection between environment and spatial policy is clearly indicated by the last point of para. 3 above, as is their common purpose in relation to economic and social cohesion. The reference to 'balanced development of its regions' in Article 130(r) is one of the additions inserted at the Maastricht Summit. Thus, like spatial policy itself, the profile of environment policy is by no means unconnected to the high politics of European integration and enlargement.

The environment title of the TEU contains only three articles. Article 130(r), quoted above, is concerned with general objectives and principles. Article 130(s) states that proposals on all except a specific list may be adopted by QMV (see Chapter 3). This article is noteworthy for the inclusion of 'measures concerning town and country planning [and] land use' among those requiring unanimity (see Figure 4.1). This is seen as a power that could be used to enhance control over actual or potentially polluting land uses, but the context within which this text is set in the treaty is such that it has not been seen as a general competence in spatial planning in spite of the terms used in the other language versions of the treaty (see Chapter 4) and the interpretation placed on it in the *Denmark 2018* study (see Chapter 6). In any event, this reference to town and country planning has not been used as a legal basis for proposed legislation and is unlikely to be until after the information in the Compendium project (see Chapter 12) is available.

Article 130(t) allows any member-state to adopt more stringent environmental standards than those laid down in an EU directive or regulation. This was included to meet Danish interests in the SEA and again in the Maastricht negotiations, and is also needed to satisfy Finland and Sweden.

Switzerland also adopts strong national environmental legislation and would normally expect to base this on principles at least as strict as those of the EU. Indeed, the rejection of the EEA agreement (see Chapter 3) by Switzerland in the 1992 referendum and the subsequent vote against contruction of lorry routes and in favour of rail routes only through the Alps reflect popular support for a strong environmentalist line.

Relationship with employment and economic development

Why should an 'economic' community be concerned with environment policy? Given the international and crossborder nature of many environmental issues and their high political salience, the principle that the EU should have environment among its competences does not cause widespead surprise or dispute nowadays, but in early statements and action programmes on the environment it was felt necessary to address this question. The argument is worth outlining here as an aid to understanding the logic and priorities of EU environment policy.

Essentially, the rationale for an environment policy in an economic body such as the EU follows from the principles agreed at the Paris Summit of 1972 (see Chapter 5), and is based on three lines of reasoning:

1. The competition or single-market argument.
2. The quality-of-life argument.
3. The economic benefit argument.

The competition argument is the fundamental reason, based on the fact that the Treaty of Rome set out to create a 'common market'. It follows from the creation of the single market, the adoption of the four freedoms, and the consequent need to remove non-tariff barriers and any form of distortion of competition owing to differential levels of environmental regulation. This is sometimes expressed by the metaphor of creating a level playing-field. Variable environmental and pollution control standards between different member-states are seen as a form of non-tariff barrier. If a situation were permitted to develop in which member-states took widely differing actions to deal with environmental problems and combat pollution, there would inevitably be distortion of competition resulting from the presence of pollution havens and cheaper operating costs, interfering with the proper functioning of the common market (Fairclough, 1983; Williams, 1986).

The quality-of-life argument can be expressed in either social or economic terms. Those emphasising the social viewpoint argue that the EU does not exist solely to benefit business and governments, that all EU citizens are entitled to enjoy the benefits of a high standard of living, and that high environmental standards are an essential part of this. The economic emphasis is based on the argument that quality-of-life criteria affect location decisions made by entrepreneurs (whether or not the industrial or commercial activity being undertaken is directly dependent on a clean environment) since the decision-makers would choose to locate – other things being equal – in cities or regions with an attractive environment offering good living conditions.

The economic benefit rationale is based on the argument that in certain sectors of the economy, the quality of the product or the attainment of satisfactory operating conditions are directly dependent on high environmental quality, and that the markets for certain manufactured products or services are stimulated by the requirement of government or the expectation of society that high environmental standards should be adopted. It is further argued that the manufacture of environmental technology products is itself a large and growing sector of the economy, providing a significant number of new jobs.

Which are the sectors of economic activity most dependent on a clean environment? Clearly, the prime categories are fisheries, forestry, tourism and agriculture, although the latter is often accused of being the source of EU-subsidised pollution as a result of runoff from chemical fertilisers entering watercourses. The bathing-water directive (EEC/76/160; Haigh, 1989: 61) has become a factor in the management and marketing of tourist resorts, bringing environment policy directly to the consumer's attention.

Secondly, it is argued that anyone concerned with local economic development, whether through inward investment and footloose industry or the location and development of business parks and commercial property development designed to attract occupiers in modern new-technology industries, must give priority to maintaining high environmental standards and offer high-quality working conditions and quality of life in order to attract high-quality firms and maintain property values. This is evident from studies of land and property markets in France, Germany and The Netherlands, for example (Acosta and Renard, 1993; Dieterich *et al.*, 1993; Needham *et al.*, 1993). Beyond the high-technology sectors of the economy, many mobile sectors of both manufacturing and service industry base locational decisions on quality-of-life criteria and a positive environmental image of an area not only for the immediate benefit of the workforce but also to attract or retain top staff and their families. Thus there is a link to the housing market as well.

Thirdly, the emission-cleansing and control-technology sector of manufacturing industry is now very big business with good growth and employment prospects.

Principles of environment policy

Certain basic principles have been laid down in successive statements of policy. These are that the polluter pays, that prevention is better than cure, and that action should take place at the appropriate level of government. The first two are now embodied in Article 130(r), para. 2 of the TEU, and the third, in the form of subsidiarity, is stated in Article 3(b) of the treaty 'In areas which do not fall within its exclusive competence, the Community shall take action, in accordance with the principle of subsidiarity, only if and in so far as the objectives of the proposed action cannot be sufficiently achieved by the Member States and can therefore . . . be better achieved by the Community' (EC Council, 1992: 13–14). The subsidiarity principle was elaborated in the guidelines agreed at the Edinburgh Summit of December 1992.

That the polluter should pay is widely accepted in principle though sometimes difficult to achieve in practice. The greatest challenge to it comes in the area of former communist countries where the polluter may no longer exist (e.g. state enterprises of the former GDR) or have no possibility of being able to pay the sums necessary to overcome the environmental legacy of years of neglect.

The principles that prevention is better than cure and the precautionary principle have been aspirations since 1973. Adoption of these principles is made explicit in Article 130, para. 2 of the Treaty. Prevention clearly makes sense although many of the earlier measures responded to the need to give priority to remedial action against existing pollution. The precautionary principle is that, whenever there is uncertainty about whether some activity will have an adverse effect on the environment, the benefit of the doubt should be given to the environment.

The logic of prevention and the precautionary principle led the Commission to seek, when formulating directives, a sufficiently early part of the decision-making process so that the need for any preventative measure required on environmental grounds could be assessed and, if necessary, imposed by condition. This logic led DG XI to focus attention on the land-use planning process since there is, in all member-states, a procedure for the authorisation of development which takes place at a sufficiently early stage in the decision-making process for any preventative measures to be specified before the development is allowed to proceed.

The environmental assessment directive of 1985 (EEC/85/337), which is integrated with the land-use planning process, is an excellent example of the application of this logic (see below).

Subsidiarity

Subsidiarity has come into prominence since the Maastricht Summit. It is a principle that is intended to apply to all policy sectors where implementation of the policy depends on national or subnational institutions. In the environment context, subsidiarity in effect means the same as the principle long established within EU environment policy that action should take place at the appropriate level of government, whether that is at the local, regional, national or supranational level. Interpretation of this is a matter for individual member-states. It is seen by some as no more than an elementary rule of good government. However, the word has been taken over by those who argue for more power to be assigned to national rather than subnational or supranational authorities.

Environment is one of the policy sectors to which subsidiarity is not only applicable but which has also been identified, following the guidelines agreed at the Edinburgh Summit, as one to which subsidiarity is to be publically and demonstrably applied. Differences between member-states in their policy towards the environment, the interpretation they apply to EU environment policy obligations, and on the political will to pursue environmental protection objectives are becoming evident. The more that subsidiarity is applied to this sector, the greater will be the divergences that can be expected. In fact, subsidiarity offers a very convenient means whereby this can be justified by politicians.

In relation to the main sectors of environment policy (water, waste, air, chemicals, noise, wildlife and countryside), there is clearly a case for arguing that those forms of pollution that occur on land, unlike water and atmospheric pollution, are matters for national rather than EU action on the basis of the subsidiarity principle. However, even in this case it is not necessarily so for reasons which follow from the competition argument outlined above – as the case of contaminated land, discussed below, illustrates.

The member-states

Member-states play a crucial but not necessarily consistent role in environment policy, a problem which may increase with increasing emphasis on subsidiarity. In every case there is a substantial body of national environmental legislation and policy reflecting each member-state's own needs and the political salience of the environment issue in each country's national politics. European law needs to be integrated with national law, and directives in particular allow scope for different interpretation in the manner in which they are incorporated into national legislation. Therefore, the activities of national and subnational government in each member-state represent an essential element of EU environment policy as a whole.

Sustainability

Article 2 of the treaty, concerned with the basic principles of the EU, sets as one of the tasks 'sustainable and non-inflationary growth respecting the environment'.

Sustainability raises philosophical and conceptual issues, although the word is also used often, like subsidiarity, for reasons of political expediency. The accepted classic definition of environmental sustainability is that first given in the Brundtland Report (1987: 8): 'to ensure that it meets the needs of the present without compromising the ability of future generations to meet their own needs.' Since the publication of that report, and the adoption of Local Agenda 21 at the UNCED environment summit in Rio in 1992, the word has become common currency in environmental discourse and policy statements.

The environment lobby

A strong environmental lobby is a major fact of life in the politics of environment policy in the EU. There are a number of European environmental networks and organisations, and in each member-state there is a very wide range of environmental pressure groups. Public support can often be mobilised for environmental issues, and the vote for green-party candidates in EP elections and in several national elections reaches quite significant percentages.

At the European level, the European Environment Bureau (EEB) is well established, having been founded in 1974. It has close links with DG XI and is in regular consultation over proposals. It has a small staff in its office in Brussels, who monitor policy developments – acting as the eyes and ears of the

EEB and on behalf of its many member organisations – and who prepare position papers and policy analysis. The staff are directed by a 14-person executive committee whose members all come from different member-states, and who are advised by a range of expert groups concerned with themes such as the natural environment, development issues, education and urban problems, tourism and the cultural heritage, as well as more technical aspects of pollution (EEB, 1994). The EEB issues a report under each EU presidency, i.e. every six months, reviewing the state of environment policy and presenting the presidency with its priorities and proposals for the next six-month period.

Over 160 organisations from 23 countries form the membership of the EEB, and this list includes most of the main environmental organisations and several with spatial planning interests in member-states. The EEB is not the only environmental body lobbying at the EU level. The Institute for European Environmental Policy, with offices in Arnhem, London, Madrid and Paris, is an independent organisation which undertakes analysis of environmental policies in Europe, seeking to increase public and official awareness of the EU dimension of environmental protection and to promote EU policy-making in this field. It is associated with the European Cultural Foundation.

Other lobbies at the EU scale include the International Council for Local Environmental Initiatives (ICLEI), based in Freiburg, the European Sustainable Cities and Towns Campaign (see Chapter 11), and networks funded under the RECITE programme and others such as Eurocities and Environet (a network for environmental cost-sharing for peripheral cities). There are also several single-issue groups such as Cities for Recycling and the Car Free Cities Club (see Chapter 11).

Policy programmes

The Commission has, since environment policy became a recognised sector in 1973, adopted the practice of setting out a multiannual environment programme in the form of successive action programmes on the environment. There have now been five of them. These are listed in Figure 10.1.

- First Environmental Action Programme 1973–6, *OJ* C 112, 20 December 1973
- Second Environmental Action Programme 1977–81, *OJ* C 139, 13 June 1977
- Third Environmental Action Programme 1982–6, *OJ* C 46, 17 February 1983
- Fourth Environmental Action Programme 1987–92, *OJ* C 328, 7 February 1987
- Fifth Environmental Action Programme 1993–2000, COM (92) 23

Figure 10.1 *Action programmes on the environment*

Although these programmes, containing reviews and statements of policy development intended for the period covered, are non-binding, many proposals have been adopted as law. Haigh (1989) listed over 200 items of European environmental legislation, and the rate of adoption has increased from around 20 p.a. in the 1980s to over 30 p.a. (Williams, 1990: 202). Many of these were minor or narrow in scope, but the pace of environmental legislation has not slackened significantly since. During the 1990s, DG XI has placed greater emphasis on working with local and regional authorities, non-governmental bodies and networks such as Eurocities, rather than in its more traditional top-down legislative style, working primarily with national governments.

The bulk of this legislation has been concerned with one of the following sectors of the environment: water, waste, air, chemicals, wildlife and countryside, and noise. The environmental assessment directive is one of the very few measures that have been explicitly cross-sectoral.

The current fifth action programme, for the period 1992–2000, is entitled *Towards Sustainability* (Commission, 1992b; and see below). This is the most comprehensive to date and is published in three volumes. It highlights as target sectors industry, energy, transport, agriculture and tourism, and develops the theme that industry should be seen as part of the solution rather than only as a source of problems, emphasising subsidiarity and the need for bottom-up policy development.

Environmental policy measures

Among the over 300 items of EU environment legislation, there are a number that have a direct or indirect effect on land-use planning. The most significant is Directive EEC/85/337 on the assessment of the effects of certain public and private projects on the environment, commonly known as the environment assessment directive, which is discussed below. First, some others are briefly referred to.

Protection of habitat

Directive EEC/79/409 (*OJ* L 103, 25.4.79; *OJ* L 233, 30.8.85) on the conservation of wild birds is one of the first to be concerned with nature conservation rather than pollution. It is primarily aimed at controlling the hunting of wild birds, particularly when migrating, and it includes in its provisions an obligation to define special protection areas in order to preserve habitat. The principle of using land-use controls to protect habitat is extended in Directive EEC/92/43 (*OJ* L 206, 22.7.92) to the conservation of wild flora and fauna. The directive required protected areas to be designated by July 1994 and stated that land-use planning policies should be developed for the management of landscape features that constitute important habitat.

Land-use designation in the guise of environmentally sensitive areas has also been introduced into the CAP as a means of integrating environmental considerations into agricultural guidance policy while providing for special forms of assistance to farmers in the designated areas. It comes within a regulation on improving the efficiency of agricultural structures, Regulation EEC/85/797 (*OJ* L 93, 30.3.85).

Contamination and waste disposal

Legislation on waste of various types forms a coherent body of legislation (Haigh, 1989; RTPI, 1994) but some proposals such as those on the landfill of waste and civil liability for damage caused by waste have encountered difficulties that have delayed or prevented adoption. It is instructive to consider some of the factors behind this.

The problem of liability in relation to contaminated land is an issue of acknowledged significance on which proposals dating from 1989 have been stalled for some time. Contaminated land, and fear of liability associated with it, is an issue of major concern to developers, investors in development and policy-makers concerned with regeneration of urban brownfield sites. Undue fear of the risk of liability for any contamination may be deterring developers from sites where contamination is suspected, even if not confirmed. The actual incidence of a legacy of contaminated land tends to occur most in the older industrialised areas in Belgium, Germany, France, Spain and the UK (RETI, 1992; Meyer *et al.*, 1995), with the absolute worst problems in the EU occurring in the former GDR, for example in the Bitterfeld area. This area was the location of one of the largest chemical engineering complexes in Europe, characterised by out-of-date technology and minimal maintenance and pollution control.

However, the extent to which public sector bodies accept risk of liability from past pollution by industrial processes and enterprises no longer in existence, and the corresponding problem of the risk to a private developer inheriting liabilities with the acquisition of land, varies greatly between member-states. In the former GDR and other parts of Germany, such as the Ruhrgebiet, private developers are protected from much or all of the risk by the various public-private partnership agencies created to promote development in these areas, whereas similar levels of freedom from risk do not necessarily prevail in the UK and elsewhere. This has the potential to be a major distortion of the single market and therefore an EU issue as even the UK's Department of the Environment acknowledged (DoE, 1994: para. 4B.20). If contaminated land is an issue capable of distorting the operations of the land development, investment and insurance markets, it is a single-market issue and therefore clearly an appropriate subject for EU action (Meyer *et al.*, 1995; Williams, 1995a).

The Commission put forward a proposal in 1989 for a draft directive on *Civil Liability for Damage caused by waste* (Commission, 1989), based, as with environmental assessment (see below) and some other environmental proposals, to some extent on experience of federal legislation in the USA. After consultations, an amended version was presented to the Council of Ministers in 1991 which raised the possibility of a European fund for compensation for damage and impairment to environment caused by waste. Although not the same as the US Superfund, the idea may have been derived from this.

Since progress towards adopting this and another proposed directive on landfill waste has been stalled by opposition from Greece, Ireland, Portugal and the UK, the Commission had to step back from its specific proposals, issuing a *Green Paper on Remedying Environmental Damage* (Commission, 1993a) based on the premiss that different systems of civil liability could lead to distortions of competition. At the same time the CoE drew up proposals for a *Convention on Civil Liability for Damage Resulting from Activities Dangerous to the Environment* (CoE, 1993). There the matter has rested as far as the EU is concerned, although it is possible that revised proposals could find a majority under QMV since enlargement. These proposals, and their relationship to US procedures, is discussed in Meyer *et al.* (1995).

The experience of these proposals has also demonstrated that contaminated land is an issue on which some common European definitions are needed, as the Commission's 1993 green paper and the report compiled by RETI argued (RETI, 1992). For example, there is scope for confusion between the exact senses of words such as the German *Altlasten* and *Brachflächen* and the distinction between them in domestic usage (the former implying greater contamination), and the exact sense in which equivalent English words such as derelict land are used. Within the UK, 'derelict land' is the term associated with government programmes over many years, while 'contaminated land' has become widely used only in this decade. In an attempt to achieve linguistic neutrality, the Euroenglish terms 'damage' and 'impairment' were proposed by the EU Commission in its draft Directive (Commission, 1989).

The environmental assessment directive

The directive on the assessment of the effects of certain public and private projects on the environment, Directive EEC/85/337 (*OJ* L 175, 5.7.85) was foreshadowed in the second action programme on the environment although it was not adopted until 1985. The original inspiration for the concept was US federal legislation – the 1969 National Environmental Policy Act requiring environmental impact statements for major projects (Wood, 1988) – although the actual EU procedures are a long way removed from those of the USA. The US legislation acquired a reputation for generating litigation over the content of environmental impact statements, a factor which explained some of the opposition to the EU directive when it was proposed.

Drafts were first prepared as early as 1978, and the formal proposal was tabled in 1980 (Commission, 1980b). Following consultations with governments and interested groups, the Commission submitted a revised proposal in 1982 (Commission, 1982). On this basis, a draft which met the objections of the less enthusiastic members of the Council of Ministers was framed by late 1983. This then proved unacceptable to the Danish government, who were reviewing their own environmental legislation and were concerned that it was too weak. Eventually, it was adopted in 1985 and entered into force on 3 July 1988. At the time, it was reputed to have been the subject of more consultation and revised drafts than any other directive on any subject, partly in order to meet French and British objections.

At that time the danger of the emergence of pollution havens creating such distortions was not fully appreciated, since the idea that one should think in terms of the EU as one jurisdiction, one market and one economic space was still not grasped by all concerned. This is well illustrated by the position taken by several organisations in the UK who opposed the then proposed Directive on Environmental Assessment of Projects on the grounds that the UK already had legislation allowing its objectives to be met, i.e. the Town and Country Planning Acts, and therefore had no need of European legislation (Williams, 1986; 1988). This argument, which is resurfacing in anticipation of 1996 IGC, completely missed the point of the idea of a common market. Overcoming this type of reasoning or misunderstanding concerning the nature of the Community was one of the purposes of the SEM programme leading up to 1992 target date.

The text of the directive is not long, consisting of 14 articles, some purely procedural, plus three annexes. Article 1 defines its scope. The term 'developer' is defined to include both private developers and public authorities responsible for projects. Development consent is defined as authorisation to proceed given by the competent authority designated by the national governments. In its effect, this means planning permission in the UK or Irish sense, the *permis de construire* in France or the *Baugenehmigung* in Germany. In a significant amendment to the original proposal, Article 1 excludes not only national defence projects but also those authorised by specific Act of the national legislature, with the justification 'since the objectives of this Directive, including that of supplying information, are achieved through the legislative process' (Article 1.5). This loop-hole is possibly one reason why authorisation of major projects by private Act of Parliament, most notably in the case of the Channel Tunnel rail link, has become more popular in the UK.

The essence of the directive is expressed in Article 2.1: 'Member States shall adopt all measures necessary to ensure that, before consent is given, projects likely to have significant effects on the environment by virtue, inter alia, of their nature, size or location are made subject to an assessment with regard to their effects.'

The text goes on to specify scope of the assessment, information requirements, public participation and consultation with other member-states. Article 4 and the first two annexes define the types of project subject to assessment. Annex I list projects which are likely to pose the greatest threats to the environment, which are suject to full assessment. These include chemical and steel works, oil refineries, power stations, nuclear installations, and major transport infrastructure such as airports, motorways, high-speed rail, ports and inland waterways. Annex II lists a much longer range of projects, many of which can occur frequently within any local or regional authority, for which a simplified form of assessment may be proposed 'where Member States consider their characteristics so require' (Article 4.2). It was supposed that this discretion extended to deciding that an assessment was not necessary at all, but subsequent legal clarification established that total exemption was not allowable.

The ways in which it has been incorporated into national law vary considerably. In the UK, this has been done by incorporation into existing legislation by means of Statutory Instrument 1199, The Town and Country Planning (Assessment of Environmental Effects) Regulations 1988, and similar regulations under highways and other Acts. At one time, it was expected that only a handful of projects annually would fall within the terms of the directive, and that for any one local or regional authority it would be a rare occurrence. In fact, it has become a relatively routine feature. A thorough introduction to the subject may be found in Glasson *et al.* (1994).

The *Green Book on the Urban Environment*

The *Green Book on the Urban Environment* (Commission, 1990b) was one of the most creative pieces of work in the spatial policy field to be produced by the Commission. It was published in the year prior to the Maastricht Summit as part of a campaign by the then Commissioner, Carlo Ripa di Meana, to develop a programme and a budget for urban environment policy within DG XI. In this objective it failed, although it did stimulate a great deal of interest and debate among urban and environmental policy-makers. However, a number of other initiatives taken in its wake have survived and are of continuing significance. These include the EAUE (see below), and the Expert Group on the Urban Environment, the Sustainable Cities campaign and the Car Free Cities Club, discussed in Chapter 11.

The green book sets out urban planning principles that would meet environmental and energy-efficient objectives, including policies to minimise transport dependency in cities and to encourage reuse of brownfield, or formerly urbanised, development sites in order to avoid urban sprawl and excessive conversion of greenfield land. It met with opposition in some member-states which might have been expected to support its environmental aims – such as Germany, because it was seen as extending beyond the competence not only of the EU (which was deliberate, with the IGC in mind) but also of the Federal government (infringing *Länder* or states' rights in respect of urban policy – Kunzmann, 1990). The green book can still be regarded as one of a number of attempts to set an agenda for an EU urban policy, and in this light it is discussed further in Chapter 11.

European Academy of the Urban Environment

One institutional outcome of the green book was the foundation of the European Academy of the Urban Environment (EAUE – Europäische Akademie für städtische Umwelt) in Berlin. This was an initiative of the Land of Berlin and was founded with some financial support from DG XI at the time when it was envisaged that the urban environment would become an established policy field within the Commission. It has been in operation since 1991 and offers short training programmes, conferences and colloquia, and undertakes studies and demonstration projects on urban traffic, urban ecology, and sustainable architecture and planning techniques in order to promote dissemination and the exchange of ideas and experiences.

Environment and transport

A green paper which followed the green book on the urban environment, but which attracted less notice and has also been a victim of the low profile of DG XI in the early 1990s, is the *Green Paper on the Impact of Transport on the Environment*, subtitled *A Community Strategy for 'Sustainable Mobility'* (Commission, 1992c). This green paper builds upon that on the urban environment and on the fourth action programme, and is an attempt to address the criticism that different DGs can find themselves pursuing contradictory policies, especially in respect of the environment.

As with the urban environment paper, it did not lead directly to a programme of legislation but it did feed directly into the preparation process of the fifth action programme and, via the white paper on the common transport policy (see Chapter 9), into debates concerning the TENs. It also identified ideas that could be incorporated into national or regional policies or on which research is needed.

Land use and the relationship between transport provision and urban form are discussed, as is the land-take of different forms of transport infrastructure. It examines the available data on lengths of road and rail networks to estimate that, in 1986, 28,949 km^2 or 1.3% of the total land area was covered by the road network, while 706 km^2 or 0.03% was covered by the rail network (Commission, 1992c: 27). As the transport white paper pointed out, the important measure of land use is the infrastructure land-take per unit of transport, expressed in persons or tonnes per unit of distance travelled. On this basis, air and sea are the most efficient, then inland waterways, rail and least of all road.

The fifth action programme on the environment

The fifth action programme on the environment (Commission, 1992b) is contained in three volumes. Volume I is the formal *Proposal for a Resolution of the EC on a Community Programme of Policy and Action in Relation to the Environment and Sustainable Development*, submitted by the Commission to the Council of Ministers. The proposal was adopted by the Environment Council in December 1992, with revisions to the text to incorporate the Council's views concerning key issues and priorities for Commission attention.

Volume II, entitled *Towards Sustainability*, is the main text of the programme, setting out the policies, proposals for EU action and responsibilities for implementation. Volume III, entitled *The State of the Environment in the European Community*, is a review of the problems being faced, and it presents the analysis and justification of the policy proposals. Five key sectors are identified in Volume III where integrated approaches to sustainable development must be adopted in order to protect the future environment: agriculture, energy, industry, transport and tourism.

The main features of Volume II that have implications for spatial policy, or which may require action by spatial planning authorities, are summarised. Air quality is a continuing concern in spite of several existing directives, especially in urban areas. Proposals to deal with air quality and acidification by setting specific targets for the reduction of nitrogen oxides and sulphur dioxide include revision to the existing directive on municipal waste incineration and a new directive on incineration of hazardous waste.

Water is the sector with the longest list of existing EU legislation, but problems still grow. Prevention of groundwater pollution by control of sources, and extensive monitoring, are proposed. Soil quality, by contrast, has been relatively neglected but is identified as a sector to be addressed. Waste disposal, especially by landfill, is a frequent cause of water and soil pollution. Problems of industrial waste disposal are rapidly becoming prominent in those parts of the EU, such as some Objective-1 and Cohesion Fund areas, which are undergoing rapid industrialisation. Land-use planning processes and environmental assessment, as well as monitoring and technical controls, will be required.

The more general issue of the quality of life is one where spatial planning is identified as playing a key role. Urban areas, although they provide the base for most economic activities and accommodate the bulk of the population, are seen as deteriorating owing to lack of investment in renewal and the abandonment of disused industrial land. Rural areas present problems arising from changes in farming methods; the need to take some land out of agricultural production and find alternative land uses; desertification; and coastal development, especially on the Mediterranean coasts. Land-use planning to achieve optimal management of growth, protection of the cultural and natural heritage, protection of open spaces, traffic management and promotion of public transport are all seen as playing a role.

Noise in urban areas is also an issue to be addressed not only by noise abatement programmes and limits on acceptable noise exposure levels but also by spatial planning of infrastructure development, industrial areas, land use around airports and other noise sources.

Transport is clearly recognised never to be environmentally neutral. Spatial planning and the integration of environmental assessment into the authorisation process are seen as critical to the reconciliation of the conflict that can clearly be expected between the EU's cohesion strategy and ambitions in relation to TENs, on the one hand, and the 'high environmental standards' on the other. In addition to spatial planning and co-ordination of transport networks, it is argued that improved planning can reduce transport needs and car dependence, while integrated public transport and development of rail and water transport for freight will also be important.

Tourism is identified as a major sector in its own right, and one whose development has yet to reach its full potential. Furthermore, the interconnection of economic and environmental factors could not be closer than it is with tourism, given its importance to the economy and its dependence on a clean environment to attract customers. Tourism along the Mediterranean coast has given rise to problems of waste management and water supply. Also the scale of urbanisation in order to maximise the number of visitors has, in some places, exceeded the capacity of the local infrastructure and natural visitor attractions to absorb the numbers. In some cases, such development has proved unsustainable as the numbers of visitors fall off in the face of these problems.

Mountain areas are another category of sensitive environment where tourist pressures can prove unsustainable and cause lasting environmental damage. This was an issue raised by Austria during enlargement negotiations, especially in the context of holiday-home development in the Alps.

Spatial policies, including strict land-use planning and traffic management, application of environmental standards, protection of sensitive areas and creation of buffer zones, as well as diversification of the economic base, are seen as part of the solution. Local and regional authorities are recognised as playing a key role through controls on development, management of tourist demand, development of pilot projects, and exchange and promotion of good practice.

Strategic environmental assessment is a concept that is referred to a number of times, and on which a directive is intended (Glasson, 1995). This concept has antecedents dating from the period when the environmental assessment directive was prepared. At that time, DG XI had the idea of proposing parallel directives for the environmental assessment of plans and programmes. Strategic environmental assessment is seen as a means whereby issues such as the quality of life, biological diversity, development of renewable energy sources and the environmental impact of industrial development can be assessed in advance of decisions being taken to authorise specific proposals.

An interim review of progress made in implementing the fifth action programme was published in 1994 (Commission, 1994f), which expressed cautious optimism regarding progress resulting from the programme. Among other recommendations, the review urged evaluation of the implications of current technological developments on physical planning and the structure of urban regions.

Information and monitoring

Monitoring environmental change and identification of trends and emerging problems impose massive information requirements, leading to the development of special systems and a new agency. The Commission has published periodic reports on the state of the environment (e.g. Commission, 1987; Stanners and Bourdeau, 1995) that are to some extent political documents taking stock of current problems and policy development, and that provide a basis on which to set priorities within the action programmes. However, these are necessarily limited in their degree of detail and often subject to problems of variable definitions and data availability.

CORINE

A system for the assembly of information on the state of the environment, known as CORINE, was devised on the basis of a decision in 1985 (Directive EEC/85/338; Haigh, 1989: 357–9), the full title of which was *Decision on the Adoption of the Commission Work Programme Concerning an Experimental*

Project for Gathering, Coordinating and Ensuring the Consistency of Information on the State of the Environment and Natural Resources in the Community. The objective then was to set up an experimental project running for four years to devise a method of collating information on four priority issues:

1. Biotopes for conservation.
2. Acid deposition.
3. Protection of the Mediterranean environment.
4. Improvement in comparability and availability of data and methods of analysing data (Haigh, 1989: 357).

Based on the experimental work undertaken, the CORINE Land Cover project has established a comprehensive database of land use and how it is changing for the whole EU by means of satellite imagery. A set of 44 classifications of land cover are used, and in combination with other databases it provides a basis not only for environmental monitoring but also other aspects of spatial policy and territorial issues. In fact, the *Europe 2000+* report (Commission, 1994a) uses examples of its mapping for its front-cover design.

Europe 2000+ identifies three broad categories of policy development which can be aided by CORINE, under the headings of territorial overview, evaluation and anticipation, and sustainable planning (Commission, 1994a: 84). The territorial overview facility will allow comparisons to be made across administrative borders on the distribution of major ecosystems and biotypes, and the spread of urbanisation or certain farming operations. CORINE will also make it possible to identify areas for which specific spatial policy measures may be required.

The observation of land-use change over a period of years enables effective evaluation of the extent to which existing policies have been implemented, and can help predict future changes and guide future policy. CORINE data showed, for example, that wetlands in western France dried at a rate of 15–20% p.a. during the 1973–91 period and identified the need for appropriate conservation strategies.

Sustainable planning is aided by data on land-cover and ecological characteristics, enabling possible adverse consequences of certain projects to be avoided. For example, by comparing data on wine-growing areas with areas at high risk from erosion, areas potentially threatened by grubbing programmes to remove old roots can be highlighted.

The European Environment Agency

The European Environment Agency finally started operations early in 1994 after long delays in agreeing its location in Copenhagen, although it did not reach its full complement of 40–50 staff until 1995. Its Board of Management consists of one representative from each member-state plus two from the Commission and two from the EP. Initially, it will be little more than a monitoring agency as it has no powers of policing and enforcement, nor of formulating environmental policy. It will take over the CORINE programme and collate statistical data on environmental quality, reporting direct to the Commission and EP. It will also publish an annual report. It is intended that it will develop assessment methodologies and set up a network of national focal points in order to provide the Commission with data and advice.

Priority topics identified in its first multiannual programme for 1994–9 include air quality and atmospheric emissions; water quality, resources and pollution; soil quality, flora, fauna and biotopes; land use and natural resources; waste management; noise emissions; chemical substances hazardous to the environment; and coastal protection. In particular, transfrontier, transnational and global aspects of these will be emphasised.

When fully operational it will provide member-states and any other countries wishing to participate with objective information on which to formulate and implement effective environmental policy; be a source of technical scientific and ecological data on which to base technical operations and legislation; provide early warning of problems requiring preventative measures; and input EU environmental data into international programmes.

After two years of operation, the Council is to consider whether to give the agency wider responsibilities. Possible new tasks could include monitoring the implementation of EU environment legislation; preparing criteria for, and awarding labels to, environmentally friendly products; promoting environmentally friendly technology; and establishing criteria for assessing environmental impact.

Compliance

The issue of compliance with environmental directives, and of enforcement of their provisions, is one that regularly presents difficulties since there is often short-term economic advantage to be gained by not observing them strictly, and the Commission does not have the resources to police the policy. One motive for establishing the European Environmental Agency was to perform this role. However, member-state governments have ensured that, initially at least, its role will be confined to monitoring and data collection, and fall short of policing and enforcing EU environment policy.

Meanwhile, considerable differences between member-states are apparent in formal compliance and the extent to which directives have been implemented. Denmark is the only member-state where the compliance rate for environment directives is better than the average rate for all directives. The other countries with good rates are Germany, The Netherlands, Spain and Portugal (although the latter two have some derogations as part of transitional accession arrangements). Italy, followed by Greece, have the worst records, and the UK is in a middle group with Belgium, France, Ireland and Luxembourg (Wurzel, 1993).

Forthcoming issues

A number of proposals and new policy initiatives of significance in spatial planning are to be expected as the new commissioner develops her programme and the new member-states begin to exert influence on the balance of opinion and voting in the Council of Ministers. Among these are proposals to revise the 1982 directive on major industrial accidents involving dangerous substances, known as the 'Seveso Directive' since it was adopted in response to the Seveso disaster and the earlier Flixborough explosion (Haigh, 1989: 254). It is likely that the revised version will include specific requirements concerning land use around establishments covered by the directive.

Other policy directions that may be anticipated include the greening of transport and agriculture policies, and giving the Structural Funds a more environmental orientation. Inclusion of environmental objectives and priorities within CSFs will be supported, as will procedures for assessment of proposals to ensure that they are environmentally friendly. The time has also come to look again at the environmental assessment directive with a view to revising it in the light of experience and extending the scope of the assessment system. The original concept that assessment would be extended to policies and programmes as well as plans is again receiving attention, although with opposition from the UK (EIS, 1994b: 25), and the concept of strategic environmental assessment can be expected to be on the agenda.

The CEE

Since 1989, a major priority has been to establish political and economic links with former Communist Party states and assist in their transformation. All have had the experience of regimes which have paid no attention to concerns over pollution, air and water quality, or conservation of the natural environment. The legacy of this period is pollution of a different order altogether from those typical of even the worst examples in the present member-states. The Görlitz–Usti–Katowice triangle in the border regions of the former GDR, Silesia in Poland and northern Bohemia in the Czech Republic, contains some of the worst polluted environments in Europe, if not the world.

These countries are now attempting to overcome this legacy, supported *inter alia* by EU aid programmes such as PHARE and TACIS (see Chapter 2), both of which include recovery from pollution and land contamination among objectives eligible for funding. The countries which are aspiring to EU membership are therefore attempting to bring their standards up to EU levels in order to be able to negotiate membership.

Conclusion

As has been shown, sound environmental standards and the adoption of higher environmental standards are intended to underlie all EU policies and should therefore be of ever-increasing salience. There has been something of a pause in the rate of growth of EU environmental legislation in the mid-1990s, coupled with recognition that regulation is not the only way to achieve environmental objectives. Engaging in dialogue with the environmental lobby and with local and regional authorities and city networks can generate pressures for improved standards and support for higher standards which DG XI would willingly respond to.

As the new European Environment Agency gains experience, it will be able to indicate where action is needed, based on EU-wide data. The 1995 enlargement has had the effect of adding environmental 'push-states' to the EU and increasing political support for the highest attainable standards. A number of proposals have been presented to the Council of Ministers which have not yet made progress, for example on issues such as strategic environmental assessment and civil liability for contaminated land, which could readily be adopted given a change in political will.

11

Towards an urban policy

The Commission has no member or DG for urban policy, nor can the subject-matter of this chapter be related to a title in the treaty or specific competence given to the Commission. It is included for three reasons. First, a body of EU urban policy does exist and is growing. Secondly, many of those professionally engaged in spatial planning in local and regional authorities also have responsibilities for urban policy although, in the eyes of DG XVI, they are two distinct policy sectors. Thirdly, there are many interest groups putting forward the argument that the forthcoming IGC should propose amendment of the treaty to insert an urban title, just as an environment title was added by the SEA, or at least create an EU competence over urban policy. Overall, considerable lobbying effort is being directed to this end with bodies such as Eurocities to the fore.

Moves to develop a coherent EU urban policy may reflect a more fundamental political imperative of redressing the dominence of the CAP. Around 80% of the EU's population lives in urban areas, yet at one time over 80% of compulsory expenditure (expenditure that is obligatory as a direct consequence of action adopted under the treaty, or approximately two-thirds of the total budget) was for farm price guarantees under the CAP (Nugent, 1991: 317). In the preliminary draft budget for 1996 (Commission, 1995a) this was down to under 48% but the disparity with an urban population remains.

Several studies have been undertaken into urban problems in the EU, and into the comparability of urban data from different member-states, in order to provide a basis for identifying issues which occur across the EU and which might appropriately be addressed by an EU policy instrument. One study undertaken for DG XVI in the 1980s was based on the most meticulous collection and adjustment of data in order to overcome the fundamental problem confronted whenever analysis of urban trends throughout the EU is attempted, that of ensuring that comparative analysis of different cities is based on comparing like with like. This was undertaken by Cheshire and Hay (1989), who developed the concept of the functional urban region for which data was adjusted so that the distorting effect of national differences of administrative boundaries and definitions were neutralised. They collected data from every city of over 330,000 population in the then EU of 12 for analysis. A set of 53 larger cities were also subject to more detailed analysis as case studies.

Later studies include a programme launched in 1994 by DG XII (Science, Research and Development) into problems faced by cities and of technological options for their resolution. Themes included new urban concepts, integration of cities into their regional context, intraurban and transurban networks and methods of evaluation of possible solutions (EIS, 1994c: 55).

There have been in a sense several urban policies. Several DGs, notably DGs V, XI, XIII, XVI and XVII, have taken initiatives at different times to introduce a more specifically urban dimension to their particular policy sector, not all of which have survived in the form originally intended. Some of the reforms of the ESF proposed in the 1980s, prior to the co-ordination of the Structural Funds around the five objectives (see Chapter 5), were intended to concentrate funding on so-called urban black-spots. This was seen at the time as a device to allow the ESF to support urban economic development initiatives in inner-city locations which could include relatively compact areas designated within otherwise prosperous cities.

Although this approach no longer operates in this form, it is a noticeable feature of the Objective-2 areas, especially in the 1994–7 period, under the Structural Funds that several of the designations are small areas focusing on problems at the urban scale. More explicitly, the basic objectives of the urban black-spots concept, with its small-area urban focus, continue to be a feature of several Community Initiatives and RECITE projects, notably the Urban Pilot Projects and the URBAN Initiative, funded under Article 10 of the ERDF regulation.

A rather more distinct approach to urban policy is represented by the *Green Book on the Urban Environment*, published by DG XI in 1990 when that DG was taking initiatives under Commissioner Ripa de Meana (Commission, 1990b). Its analysis and agenda is outlined below in the first part of this chapter. The Community Initiatives are discussed next, followed by the elements of urban policy to be found within the energy and transport sectors. Finally, the chapter raises the question of the prospects for an urban policy at the IGC.

Urban environment policy

A concerted effort was made by DG XI in the period coming up to the Maastricht Summit in 1991 to establish its claim to take a leading role within the Commission on EU urban policy, as Chapter 5 indicated. The main foundation for this claim was the *Green Book on the Urban Environment* (see below). Although the ambitions of DG XI were not fulfilled, urban environmental issues continue to be of great concern to municipalities and environmental networks throughout the EU, supporting a number of other legacies of this phase of policy-making such as the Car Free Cities Club and the Sustainable Cities project (see below).

Green Book on the Urban Environment

The circumstances surrounding the production of this document (Commission, 1990b) are described above in Chapters 5 and 10. It is often referred to as the

Green Book although its correct title is *Green Paper on the Urban Environment* and its formal status is that of a *Communication from the Commission to the Council and Parliament.* Although the urban environment initiative of DG XI no longer has any momentum attaching to it as far as the Commission is concerned, the fact remains that the *Green Book* did catch the imagination of many professionals and politicians in the urban planning community, and its ideas continue to be taken seriously as a contribution to planning thought (RTPI, 1994).

It could be seen equally as a contribution to environment policy or to urban policy. The themes and contents are outlined here because it sets out one possible agenda for an EU urban policy. The first part of the document considers the role of cities in society and the essence of European urbanisation, going on to examine the urban environment and the root causes of urban degradation. The second part seeks to develop an EU strategy for the urban environment and identifies a series of areas for action, several of which clearly fall within the scope of spatial planning.

The introduction to the subject-matter is rather philosophical, tracing the concept of the city and its place in European civilisation from the Italian city-states, the Hanseatic League and the great university cities such as Bologna, Coimbra and Prague to the Leipzig demonstration of 1989 which effectively undermined the authority of the GDR. This comes across rather strangely and may be off-putting to many urban planners and policy-makers of a strictly practical orientation. The aim is to establish the central place held by cities in European culture from ancient times, and to argue that in a continent where approximately 80% of the population are urban or suburban residents, quality of life and cultural identity are crucially dependent on the quality of the urban environment.

A number of specific connections are made to areas of EU concern. The commitment to social cohesion is seen as having an urban dimension as problems of rural poverty often manifest themselves in urban areas as a result of migration. More directly, the connection between urban development and air pollution is analysed at considerable length. Other urban planning themes such as conservation of the cultural heritage, protection of historic buildings and of green spaces, traffic calming, pressures of suburbanisation and tourist development are also reviewed.

A theme that runs through the first part of the *Green Book* is that of the limits of urban areas and the contradiction between the classical concept of a European city and the extensive suburbanisation of human settlements supported by non-rural economic activities without being truly urban in their form.

Proposals for an urban environment policy

The second part of the *Green Book* sets out goals and guidelines for an EU urban environment policy, and the legal framework on which it could be based, which at that time had to be found in the environment title of the SEA. It then works systematically through the various component themes proposing suggested lines of EU action.

The overall objectives of an urban environment policy are stated as 'the creation, or re-creation, of towns and cities which provide an attractive environment for their inhabitants, and the reduction of the city's contribution to global pollution' (Commission, 1990b: 31). The two causal factors to be addressed are identified as the uncontrolled pressure placed on the environment by activities which normally are concentrated in urban areas, and the 'not unrelated factor' of 'the spatial arrangement of our urban areas'. Policies for the latter which have become widely accepted throughout Europe as good planning since the Charter of Athens of 1928, namely, of segregation of land uses, are seen as being responsible for many of the problems associated with spatial arrangement, especially when land-use zones are not well connected by public transport – so creating a dependency on private vehicles.

The principles of policy co-ordination, sustainability and subsidiarity are invoked, and the existing policy framework including CSFs under the Structural Funds, research programmes such as DRIVE and Urban Pilot Projects (see below), before specific proposals for action are made.

The areas for action identified fall into two categories. Under the heading of policies for the physical structure of the city are

- urban planning;
- urban transport;
- protection and enhancement of the historical heritage; and
- protection and enhancement of the natural areas within cities.

The second category is policies to reduce the impact of urban activities on the environment, which address

- urban industry
- urban energy management
- management of urban waste
- water management.

Within the broad remit of urban spatial policy, a number of principles were put forward, some of which challenge orthodox practice in several countries. Mixed urban land uses and higher densities (so that the car can become an option), reuse of abandoned and derelict industrial land with public intervention to overcome problems of contaminated land, revitalisation of social housing estates, and encouragement of architectural innovation were proposed. It was suggested that the Commission, in co-operation with member-states, should evolve guidelines for the incorporation of environmental factors into town planning strategies, with the objective of influencing planning practice in a direction complementary to the then proposed directive on the environmental assessment of policies, plans and programmes. The latter is now on the agenda under the title of strategic environmental assessment, while some of the other themes have been addressed by Urban Pilot Projects.

The many consequences of the growth in private car ownership and of road freight transport are an important theme which was also addressed in the 1992

Green Paper on the Impact of Transport on the Environment (Commission, 1992c; see Chapter 10).

Perhaps because of the rhetorical style in which the contextual material is presented, the *Green Book* was dismissed in some quarters as impractical nostalgia for a preindustrial city. A more serious criticism was that the proposals for action encroached on matters over which the EU had no competence, and which in some member-states were primarily or exclusively matters for regional and local government. This point of view was most strongly expressed in Germany, a country which itself has highly developed urban environment policies, because of sensitivity over the legislative independence of the *Länder*.

One of Commissioner Ripa de Meana's objectives in promoting the *Green Book* enthusiastically in 1990 was to secure a budget line for the urban environment allocated to DG XI. When it became clear that he had not succeeded in this objective, his enthusiasm for the *Green Book* diminished and it became even more marginalised after he left the Commission. Nevertheless, many of the ideas and critiques of orthodox urban planning policy have reappeared in other policy contexts, or are being promoted in other ways at arm's length from the Commission by bodies such as the Car Free Cities Club or the European Environment Bureau.

Europe 2000+ introduces a number of ideas from the *Green Book* in its section on urban areas, especially in the context of 'an integrated approach to sustainable development' (Commission, 1994a: 105–6). Its ideas also find a common focus on urban transport with some of those developed by other DGs, for example out of the DRIVE programme of DG XIII (see below).

The legacy of the Green Book

One institutional outcome of the interest in the urban environment was the establishment of the European Academy for the Urban Environment (EAUE) in Berlin in 1991, referred to in Chapter 10.

Another legacy is the Urban Environment Expert Group, formed by the Council of Ministers in 1991 to follow publication of the *Green Book*. This consists of representatives from each member-state, including six local government experts from cities of varying sizes located in different parts of the EU. It was sponsored by DG XI, who initially chaired it although that role is now taken by one of the expert members. Other DGs participate, along with the European Foundation for Living and Working Conditions, OECD, the CoE and the World Health Organisation. Support services are provided by consultants and the EAUE.

The expert group was at first engaged in advising on a wide range of policy issues and proposed legislation, but after the departure of Commissioner Ripa di Meana, the loss of Commission support for the urban environment programme caused the group to suffer a loss of momentum. A process of rethinking its role led to the agreement to focus its attention on the theme of sustainable development in cities, under the title of the Sustainable Cities Project (Fudge, 1993). Working groups drawn from the membership of the expert group examined a range of issues arising out of EU and other statements, including the *Fifth Action Programme on the Environment* (Commission, 1992b), the UNCED's Local Agenda 21 and the OECD Ecological City project, and reviewed national urban planning experience in order to be able to compile a guide to best practice. This project now involves a network of over 40 experts with all 15 member-states and bodies such as the European Foundation, Eurocities and ICLEI represented, together with advisers from the relevant DGs.

A major conference arising out of this work on sustainable cities was held in Aalborg, Denmark, in May 1994 at the initiative of the municipality. The overall objective was to create a network of European cities pledged through signing a charter to work towards sustainable urban development. At this event, the expert group's first report on sustainable development was presented. This focused on three sets of issues relating to the urban environment (the urban economy, land-use planning and mobility), reviewed the development of thinking about sustainable cities and tried to apply concepts of systems ecology to the task of developing policy and management tools. The conference adopted a *Charter of European Sustainable Cities and Towns*, which municipalities and individuals were invited to sign, thereby becoming members of the Sustainable Cities and Towns Campaign. This has an office in Brussels and, like the Car Free Cities Club, discussed below, can be regarded as a part of the European urban environment lobby (EIS, 1994d).

Car Free Cities Club

In 1992, the then Commissioner for the Environment introduced the idea of creating a network of car-free cities during a press conference. The main aim was to open up a debate on the everyday traffic problems faced by cities and to establish the means whereby experience of controlling traffic in order to create a better urban environment could be more easily exchanged. Following approval of the fifth action programme, the urban environment unit of DG XI developed the concept and agreed with Eurocities that it should be run on behalf of DG XI from the latter's secretariat by a specially appointed coordinator.

The Car Free Cities Club (CFC), or Villes sans Voitures, was launched at a conference in Amsterdam in March 1994. Membership is sought from municipalities who are prepared to make a political commitment to the ideals it represents by signing a charter. Membership is open to all cities, whether or not they have implemented policies to calm traffic or remove cars from city centres, so long as they are prepared to work for CFC's ideals.

CFC is intended to become a networking organisation that encourages and promotes its objectives by the sharing of ideas and experience and transfer of advice from cities that have successfully developed strategies for removing excessive car traffic to those that are still to embark on this. This is to be achieved by promoting practical projects demonstrating the wide range of possible solutions to the problems of urban traffic, and by mutual encouragement and group pressure. It will act as a lobby not only on DG XI (which positively welcomes this pressure on itself) but also on other DGs, national and regional governments and political bodies that are more sympathetic to the vehicle users' lobby.

ERDF Community Initiatives

There is a very strong urban orientation to the pattern of projects supported by the ERDF under Article 10 of the ERDF regulation, in the form of Urban Pilot Projects, other RECITE programmes and the URBAN Community Initiative.

Urban Pilot Projects

The title Urban Pilot Projects is given to a series of projects designed to test out new ideas for the delivery of urban policy, cofinanced by the Commission and directed by DG XVI. In the first phase there were 21 projects, and eventually 32 went ahead. Another phase is due to go ahead in the period 1995–9. With an eye to the subsidiarity principle, the Commission does not consider that it should tackle all the problems of urban areas, and accepts that the bulk of urban intervention remains a matter for member-states.

Nevertheless, just as was the case with the *Green Book* intitiative, the Commission is aware of common issues affecting cities in different parts of the EU and is seeking to identify and promote ideas on how to improve the effectiveness of urban policy, and to facilitate the exchange of ideas between cities and governments of the different member-states.

Proposals for Urban Pilot Projects were submitted by member-states for selection on the basis of four main principles. Projects

1. should address a theme of urban planning or regeneration of European interest;
2. must be innovatory in character and offer new approaches;
3. should have clear demonstration potential so that lessons can be transferred to other cities; and
4. should also contribute to the development of the region in which the city is located.

As is always politically necessary with any such Commission programme, a balanced distribution between the member-states was also an unavoidable selection criterion. Projects were funded in all member-states except Luxembourg.

Pilot projects were not intended to deal with all the problems that occur in the selected city, nor that all possible problems should be included in the programme. It was more important to concentrate effort where it could be most effective. Often, this meant that a small sector of a city was designated, where the project could lead to the development of a solution to a clearly defined problem. Consistent with the EU as an 'economic' community with no competence in housing or welfare, the focus is on projects which could contribute to the economic functioning of the city. The range of forms which an economic development contribution could take reflects the relationship between the environmental and economic objectives of the EU discussed in Chapter 10.

Three main categories of project were initially supported, relating to the three main themes where it was anticipated that policy intervention could prove beneficial in economic development terms. The fourth was added to the programme in 1992.

The four themes are as follows (see Figure 11.1):

1. Economic development projects in areas with social problems, such as peripheral and some inner-city residential areas with high unemployment and a low level of access to skills and training.
2. Areas where environmental actions can be linked to economic goals.
3. Projects for the revitalisation of historic centres, to restore economic and commercial life to areas where the urban fabric has been allowed to decay.
4. Exploitation of the technological assets of cities.

Evaluation of the Urban Pilot Projects is proceeding, but an interim assessment pointed to many positive features (RECITE, 1995a). The ERDF funding has enabled local projects to go ahead that otherwise would not have done

Economic development in areas with social problems:

Aalborg (DK), Antwerp (B), Bilbao (E), Bremen (D), Brussels (B), Copenhagen (DK), Dresden (D), Groningen (NL), Liège (B), London (UK), Lyon (F), Marseilles (F), Paisley (UK), Rotterdam (NL)

(14)

Environmental actions linked to economic goals:

Athens (GR), Belfast (UK), Gibraltar, Madrid (E), Neunkirchen (D), Stoke-on-Trent (UK)

(6)

Revitalisation of historic centres:

Berlin (D), Dublin (IRL), Cork (IRL), Genoa (I), Lisbon (P), Porto (P), Thessaloniki (GR)

(7)

Exploitation of the technological assets of cities:

Bordeaux (F), Montpellier (F), Toulouse (F), Valladolid (E), Venice (I)

(5)

Total: 32 projects

Source
RECITE, 1994d.

Figure 11.1 *Urban Pilot Projects*

Work under way in the Porto Urban Pilot Project, Portugal. (photo: the author)

and, although the EU financial contribution is modest in relation to the total public investment, it has proved valuable. Employment and enterprises are being generated, effective local partnerships are being created to respond to vocational training needs and achieve implementation, and redundant land and buildings are being brought into beneficial use. The need to account for the EU aid has encouraged adoption of a systematic approach to monitoring and reporting. The process of implementation has often been innovative, and local teams have benefited from a much higher level of interaction and exchange of experience with other projects than would normally be the case. Above all, the projects have identified ways in which 'good practice' in turning around the fortunes of areas of urban decline can be promoted (RECITE, 1995b).

Some specific issues concerning land-use planning and property ownership occurred in a number of projects. Usually, this took the form of delays caused by difficulties meeting planning requirements, getting agreement of land and property owners, or of land assembly where ownership is fragmented. Clearly, integration of such projects with the local land-use or spatial planning process is an essential feature of any EU programme such as this, and could possibly therefore be a consideration for the Commission after the *Compendium of Spatial Planning Systems* (see Chapter 12) is complete.

Exchange of experience is always a slow process until a good understanding is reached of how to apply the lessons to different contexts. Most projects, and all the later ones, had a budget allocation specifically for dissemination because a high priority is given to the demonstration value of the projects. Some of the projects were linked to existing networks through membership of Eurocities or Quartiers en Crise. Several of the cities are located in Objective 1 or 2 areas, where action can be identified which could be added to the measures included in CSFs under ERDF.

Several pilot cities are outside Objective 1 and 2 areas. In these cases, the experience has shown that equally severe problems can occur at the urban scale, and possible policy responses have been identified. Lessons from the project will no doubt be applied to the future operations of the Structural Funds as well as to any other amendments proposed at the IGC, such as the possibility of an EU urban policy.

Final evaluation of the first phase is not yet possible, and in fact 23 of the 32 projects requested additional time to complete their programme. However, the second interim evaluation did identify possible themes for further EU policy development including urban transport, measures to combat functional obsolescence and strengthen the planning of unplanned areas, and enhancing the potential of medium-sized cities (RECITE, 1995a). It also indicates ways in which the URBAN initiative and Community Initiatives in energy, telecommunications and environment may benefit from the experience of the Urban Pilot Projects.

The URBAN initiative

URBAN is one of the Community Initiatives supported by the ERDF within the framework of the Structural Funds (see Chapter 8). It was launched in March 1994 in order to extend and improve co-ordination of EU contributions to the implementation of various urban policies drawn up at national or regional level through CSFs, and through measures under Article 10 for Urban Pilot Projects and co-operation networks under the RECITE programme.

URBAN is intended to help overcome the serious social problems in many depressed urban areas by supporting schemes for economic and social revitalisation, renovation of infrastructure and environmental improvement. Individual projects would normally run for four years and have a demonstrative character with the potential to offer lessons capable of being transferred to other cities and member-states. Priority in selection was also given to innovative projects forming part of a long-term strategy for urban regeneration.

The Commission's initial intention was to part-finance around 50 projects in clearly defined geographical or administrative areas. These were expected to be in conurbations with over 100,000 inhabitants, preferably in Objective-1 areas, in cities suffering from high rates of unemployment, decaying urban fabric, poor housing and a lack of social facilities. As an exception, smaller cities suffering from overall economic decline could be included. Two-thirds of the total budget of 600 mecus is to go to Objective-1 locations, and most but not necessarily all of the remainder goes to Objective-2 locations.

Eligible activities include launching of new economic activities such as employment projects, health and safety facilities and infrastructure and environmental improvements linked to these.

Each member-state was invited to submit its proposals for cities where projects under the URBAN programme might take place by November 1994. Integration between URBAN funding and other EU programmes including Leonardo (see Appendix II) and EIB loans was also encouraged.

Urban energy policy

Another direction from which an urban policy may come is that of the energy sector – as one of the ways in which energy policy could contribute to the overall Maastricht objectives of economic and social cohesion. A communication addressing this theme was issued by the Commission (Commission, 1993c) in order to stimulate debate on the links among economic development, social cohesion and the availability of energy resources. In July 1994, the Commission followed this by inviting tenders for projects to investigate ways in which the economic development and environmental protection of regions and urban areas may be supported by improved energy efficiency.

Directions in which EU policy development may be encouraged to extend included: the interconnection of energy networks to assist crossborder integration and the islands and remote periphery of the EU; the development of initiatives on energy and the urban environment to improve the management of urban energy networks for gas, electricity and heating; addressing problems of economic convergence, conservation and sustainable development; and efficiency of urban transport networks. The communication also urged that the experience of existing programmes such as JOULE, SAVE and THERMIE should be taken into account.

JOULE (Joint Opportunities for Unconventional or Long-Term Energy supply) was a research and development programme to develop technologies that take into account new and renewable energy sources and to contribute to environmental protection through reduction of carbon dioxide emissions. SAVE (Specific Action of Vigorous Energy Efficiency) was a programme of energy conservation measures in transport and other sectors. THERMIE (Promotion of European Energy Technologies) was a programme to promote innovative energy technologies which included applications to urban transport (see below).

One conclusion from the programme of work to support regional and local energy studies and the promotion of efficient energy planning undertaken by the Directorate-General for Energy (DG XVII) during the 1980s was that more effective implementation of the recommended forms of best practice might be achieved if responsibility was decentralised to the local level of urban authorities. Therefore DG XVII set up a programme known as CITIES designed to demonstrate the potential and range of possible applications and benefits of an integrated urban energy and environment policy in terms of the general objective of sustainable urban development.

CITIES in another ingenious acronym standing for the Community Integrated Task for the Improvement of Energy-environment Systems in cities. The aim of CITIES was to identify interesting examples of successful urban energy-environment policies, analyse the key factors in their success and provide a basis for dissemination of this information to other cities. Twelve cities were selected for the project, one from each of the then 12 member-states. The cities selected were Amsterdam (NL), Besançon (F), Braganca (P), Cadiz (E), Dublin (IRL), Esch/Alzette (L), Gent (B), Mannheim (D), Newcastle upon Tyne (UK), Odense (DK), Thessaloniki (GR) and Torino (I). The main elements of the findings of these case studies are presented in Nijkamp and Perrels (1994).

Nijkamp and Perrels conclude that there is a need for a European programme of urban energy planning in order to disseminate and encourage adoption of the innovative ideas that have been developed. One of the key factors limiting the potential of cities to develop integrated urban energy-environment policies in a way that moves towards greater sustainability is the institutional structures of urban government and the energy-supplying authorities. Another is the relationship between this policy sector and other sectors of EU policy, especially on the environment and the Structural Funds.

If an EU urban energy policy is to be developed from the experience of projects such as CITIES, it will need to develop instruments to encourage spatial planning and energy policies which achieve sustainability objectives. A form of subsidiarity will be required, allowing urban authorities the freedom to develop local policies in such a manner as to achieve the most appropriate form of integration for their own circumstances within an overall framework of EU policy that allows the Structural Funds and other programmes to be applied to the same ends.

Urban transport

A common theme in several of the approaches to urban policy developed from the environment perspective and in *Europe 2000+* is that of minimising urban transport demand through spatial planning measures; promotion of transport management aimed at reducing fuel consumption; and the impact of urban traffic by controls on road use and the integration of public transport. Urban issues also figured strongly in the white paper on *The Future Development of the Common Transport Policy* (Commission, 1992d; see Chapter 9) of DG VII.

The THERMIE programme of DG XVII was developed to promote new energy technologies, renewable energy and their application. It supports over

130 projects. Three were funded to test the application and implementation of ideas through demonstration schemes in the urban transport sector: JUPITER, ENTRANCE and ANTARES. JUPITER sought ways of achieving energy savings and reducing private transport use through a range of integrated public transport policies and the use of more fuel-efficient vehicles, based on projects in Aalborg, Bilbao, Florence, Gent and Liverpool to demonstrate their potential. ENTRANCE was concerned with the implementation of advanced telematics applications to traffic control and management in Cologne, Evora, Piraeus, Portsmouth, Santiago and Southampton. ANTARES sought to reduce congestion and achieve a shift towards use of public transport in Barcelona, Bologna, Dublin, Leipzig and Toulouse.

The DRIVE (Dedicated Road Infrastructure for Vehicle safety in Europe) programme, funded by DG XIII (Telecoms, Information Market and Exploitation of Research), is an important contribution to urban transport research. DRIVE is a programme in which 31 cities are involved in projects to develop operational tools and strategies for demand management in order to reduce congestion and the environmental impact of road transport, and improve road safety and efficient use of road space through the application of transport telematics. DRIVE I focused on the development of advanced transport telematics, and DRIVE II on trials and demonstration projects.

Others

The sectors of urban policy reviewed above do not exhaust all the elements of EU urban policy that already exist. DG I (External Affairs) funds projects under the MED-URBS programme to assist cities in the 11 Mediterranean countries associated with the EU to apply new ideas and acquire expertise. DG V (Employment, Industrial Relations and Social Affairs) is involved in urban affairs primarily through the application of Objectives 3 and 4 of the Structural Funds and Community Initiatives such as HORIZON (now part of EMPLOYMENT; see Chapter 7) and POVERTY, concerned with problems of social exclusion, under which 80% of expenditure is directed at urban areas.

An urban competence?

Clearly, the EU is active in urban policy in many different ways. The core focal point for this remains DG XVI, in spite of the attempts by DG XI to take the initiative on the urban environment. On the basis of the urban Community Initiatives funded by the Structural Funds, it was proposed at the Maastricht IGC by the Commissioner, Bruce Millan, that the legal basis for Objective 2 should include the phrase 'assistance to industrial and urban areas in decline', but 'and urban areas' was not agreed (Hughes, 1994). This would have brought urban policy explicitly within the scope of the ERDF.

Since 1991, urban actions by the EU have increased, especially in the form of Urban Pilot Projects and the URBAN Community Initiative, and from January 1995 there has been an Urban Policy Unit in DG XVI. Meanwhile, the urban lobby is very highly mobilised to press for an urban title, or at least a competence, to be added to the treaty in the 1996 IGC. Eurocities is taking a leading role in this, as is the Committee of the Regions which has adopted a report from its Urban Policy Commission, calling for an EU urban policy.

On the other hand, the subsidiarity argument is deployed by those opposed to an extension of EU competence into urban policy. Counterarguments can be expected to be put strongly at the IGC, namely, that urban affairs should remain the responsibility of local and regional authorities under overall direction by member-states (the subsidiarity argument); and from the federal governments that since urban policy is not within the competence of some national governments, they as participants in the IGC have no authority to give power over urban policy to the EU (the states' rights argument).

12

Europe 2000+ *and the state of policy development*

Having reviewed the main components of EU spatial policy sector by sector, it may be helpful to take stock of the initiatives that are underway to formulate a spatial policy for the EU and assess the point it has arrived at in the year before the IGC. The year 1995 was the year in which the *Europe 2000+* document was made widely available, and in which work on the *European Spatial Development Perspective* (*ESDP*) moved towards draft policy ideas. However, these are no more than staging points in the development of EU spatial policy, the fulfilment of which awaits the deliberations of the 1996 IGC. This chapter reviews the main elements of policy development at this interim, pre-IGC, stage, while Chapter 15 discusses future issues and prospects.

The *Europe 2000+* report was not disseminated until after the Strasbourg Informal Council in March 1995, although its official date of publication is 1994 since political responsibility was taken by Commissioner Bruce Millan, whose term of office ended at the end of 1994. It is of course printed in all 11 official EU languages and has been widely distributed to professional and academic bodies concerned with spatial planning, national, local and regional governments and to many other non-governmental and lobbying organisations with an interest in spatial issues. Many of these organisations are preparing responses and critiques, and using it either as a context for their own spatial planning reports and policy development or in order to prepare positions and briefing in anticipation of the 1996 IGC.

Meanwhile, the Committee of Spatial Development is at work preparing the *ESDP*. In terms of traditional British structure plan methodology, if *Europe 2000+* can be regarded as a report or survey, then the *ESDP* is intended to set out the policies that form the plan. A number of other initiatives have been proposed or are being implemented. Among them are the *Compendium of Spatial Planning Systems and Policies*, the Co-operation Network of Spatial Research Institutes and the concept of a *European Planning Atlas*.

Within the framework of spatial policy development and planning studies by national and regional authorities, policy suggestions and proposals for the EU scale are being developed in several countries. The examples of the Benelux Union, Denmark and Germany outlined in Chapter 6 are not the only ones.

Work on these lines is underway in the French national spatial planning agency DATAR and by many regional and municipal authorities. A prominent example of the latter is the *Barcelona 2000* study.

Europe 2000+

The development of spatial policy leading up to the preparation of the *Europe 2000+* report was outlined in Chapter 5, and much of the material in Chapters 6–11 draws upon it or draws upon the themes covered. Here, it is useful to summarise some of the key themes and note how the document has been received, and to consider what platform it provides, with the other studies and policy initiatives that are current, for the *ESDP* and the IGC.

The overall aim is to support policy development which will encourage the balanced development of European territory. Its targets are, one the one hand, the Commission itself, and on the other hand the many national, regional and local authorities and individual policy-makers who shape spatial policy throughout the EU. It must be seen to some extent as a rhetorical document making a political statement, addressed to the whole Commission and the other EU institutions advocating a particular form of policy development which recognises the spatial dimension of a very wide range of EU policy sectors. It seeks to demonstrate that the spatial dimension, just as much as the financial or environmental, is a matter for policy-makers in all the sectors identified and therefore it is not an issue which is of concern only to DG XVI and no commissioner except the Commissioner for Regional Policy.

The three themes of competitiveness, viability and cohesion are pursued in support of the overall aim of balanced regional development. Competitiveness implies not only new infrastructure but also policies to reduce the tendency towards excessive growth of major conurbations and to encourage local development potential in all regions. Viability introduces the concept of sustainable development to the economy, transport, the urban environment and reform of agricultural policies. Cohesion expresses concern with the economic and social cohesion or solidarity of the EU as a whole by seeking policies that address issues of economic and geographical disparities, peripherality and spatial and social exclusion.

These are linked to four specific policy themes which have been discussed in earlier chapters in the context of the existing state of policy development: crossborder co-operation, reducing the isolation of peripheral regions, balanced development of the urban system and preservation of the wealth of rural areas.

Political reaction

Europe 2000+ has received a warm welcome in many quarters and has been strongly endorsed by the EP and CoR at plenary sessions in June and July 1995. In his report on *Europe 2000+*, the rapporteur appointed by the Committee on Regional Policy of the EP emphasised that 'regional planning must be viewed above all as an instrument for combating disparities between the Union's various regions' (EP, 1995: 6). The report goes on to emphasise that co-ordination of all sectors of spatial policy, including TENs and environment improvements (as well as those falling within the responsibility of DG XVI), is essential.

The rapporteur for the CoR supports very strongly the approach of *Europe 2000+*, emphasising particularly the importance of working with regional and local authorities in order to achieve the policy objectives arising out of the study. The CoR was supportive of developing the spatial policy of the EU, linking this to the need for a stronger urban policy focus (CoR, 1995a).

Strengths and weaknesses

Its strengths are that it is comprehensive and integrative in its scope, rather than narrowly sectoral, recognising that the EU has developed a large body of both explicit and implicit spatial policy in several of its policy sectors and pulling together all these different elements of the EU's spatial policy (Albrechts, 1995). It provides a context for the many situations where local spatial planning policy needs to relate to the wider EU context, and it attempts to apply the concepts of environmental sustainability and equity among social groups, localities and generations to its thinking. Its thinking is international, going beyond the limits of the EU, although the EP was concerned about the need to anticipate more fully the spatial policy implications of the possible accession of Cyprus, Malta and the central European countries.

The main weakness is its lack of legal authority. It has been endorsed by the Commission as a whole, which did not happen in 1991 with *Europe 2000*, so it does have greater standing than the earlier study in relation to other Commission policies. However, it is not directly linked to their financial programming.

The treatment of subsidiarity and the proper roles of the Commission and/or local and regional authorities in spatial policy is rather ambivalent. On the one hand, the political realities of lack of EU competence in spatial policy, plus the political pressures for subsidiarity, are fully recognised, as they are in the political principles for the ESDP, quoted below. On the other hand, there is an implication that spatial policy coherence cannot be achieved without some direction from the Commission that is enforceable. There is no question of proposing a common local or regional planning system for the EU, but some overall policy parameters are not ruled out.

A number of sections of the report can be criticised for appearing to assert, rather than show from research evidence, the spatial significance of certain trends, while other factors affecting spatial change, such as increased application of telematics, may not seem to those working in this field to be fully appreciated. It would not be difficult to find grounds on which to base an academic critique, but it must be recognised that the *Europe 2000+* is first and foremost a political document, asserting the political case for policy co-ordination between DGs and between the spatial planning authorities at all levels in member-states.

The Leipzig Council

The meeting of the Informal Council of Ministers responsible for spatial planning in Leipzig in September 1994 was devoted entirely to spatial policy and to consideration of *Europe 2000+*, which was then in near final-draft form. The German government published a booklet in German, French and English setting out their presidency's ideas and proposals, together with the conclusions issued by the presidency following the meeting (BMBau, 1994a). This was widely distributed. Under the title *Principles for a European Spatial Development Policy*, these ideas were republished in April 1995 (BMBau, 1995).

The Leipzig Council took certain decisions of significance for the development of EU spatial policy. First, they instructed the Committee of Spatial Development to submit a first draft of the ESDP in 1995. Secondly, they accepted a proposal from the Commission that a network of spatial research establishments should be organised to form a European Observatory of spatial change and planning. Thirdly, it was decided that relations in the field of spatial planning with the countries of the CEE should be intensified.

ESDP

This concept dates back to the first of the Informal Councils of Ministers of Spatial Planning in Liège in November 1993, when member-states agreed to work together to develop an *ESDP*. This was based on the work already published as *Europe 2000* and was initially undertaken in parallel with the production of *Europe 2000+*, sharing some of the analysis assembled by DG XVI in preparation for the latter. The leading role is taken by the CSD, enabling member-states to work together to establish a common frame of reference to which policy-makers at all levels of government could refer. At the subsequent Informal Council in Corfu in June 1994, political principles for the *ESDP* were adopted. These were that it should be non-binding and be developed gradually in order to assist decision-making by offering more coherence at the EU scale for all policies with a spatial or territorial impact.

Certain political principles which should underlie the *ESDP* were put to the Leipzig meeting. These were that

- spatial development can contribute in a decisive way to the achievement of the goal of economic and social cohesion,
- the existing competences of the responsible institutions for Community policies remain unchanged; the ESDP may contribute to the implementation of Community policies which have a territorial impact, but without constraining the responsible institutions in exercising their responsibilities,
- the central aim will be to achieve sustainable and balanced development,
- it will be prepared respecting existing institutions and will be non-binding on Member States,
- it will respect the principle of subsidiarity,
- each country will take it forward according to the extent it wishes to take account of European spatial aspects in its national policies.

(BMBau, 1995: 37)

The same document sets out a number of operational objectives for the *ESDP* to achieve a balanced spatial structure and economic development in ways that respect the diversity of regional identities and the natural and cultural heritage.

The three overall objectives are: 1) the development objective of strengthening the economic base of structurally weak regions; 2) the balancing objective, to achieve greater equality of living and working conditions throughout the EU; and 3) the protection objective, to preserve and re-establish the diversity of landscape, natural environment, habitats, recreation areas and the cultural identity of different localities (BMBau, 1994; 1995).

The main section of the booklet published for the Leipzig Council, which sets out and discusses these objectives, is entitled *Principles for a European Spatial Development Policy* (BMBau, 1994b), and the word 'policy' is retained in the reissue referred to above (BMBau, 1995).

The use of the word 'perspective' in the English title downplays the aims of policy prescription which are held by those who advocate a more coherent spatial policy for the EU. It is interesting to note that the French title, *Schéma de développement de l'éspace communautaire* (*SDEC*), uses the word *schéma*, which was translated to the approval of the UK government in the context of the CoE project in the 1980s (see Chapter 5) as 'strategy'.

Subsequent work on the *ESDP* involves the development of three trend scenarios, looking forward to 2010, for the urban system; infrastructure; transport and telecommunications; and the natural environment and cultural heritage. These are intended to help evaluate the operational objectives agreed at the Leipzig meeting and identify the nature of the policy content that needs to be developed. The next stage will involve the generation of policy scenarios based on the analysis of the trend scenarios. The intention is that this work should be accomplished, and the *ESDP* prepared, by the beginning of the Italian presidency in 1996, although it is unlikely that this will have been achieved.

The *ESDP* is intended to set out a common understanding between the Commission and the member-states on what an EU spatial policy should address – identifying EU-scale issues, and placing the needs and interests of the member-states in the EU context on the basis of acceptance that it is no longer possible for member-states to undertake effective spatial planning in isolation. It is also intended to take policy prescription a stage further than *Europe 2000+* by identifying what actions the EU should take to promote more balanced spatial development and achieve its treaty objectives.

If the next phase of EU spatial policy development is to be more incisive, the Commission will need to be in a position to propose more specific and detailed directives or regulations. Two initiatives which will provide a support system for such tasks are the Co-operation Network or Observatory of Spatial Research Institutes, agreed at the Informal Councils at Leipzig and again in Strasbourg in March 1995, and the *Compendium of Spatial Planning Systems and Policies*, already underway and discussed below.

EU Co-operation Network of Spatial Research Institutes

This initiative of the Commission, agreed at Leipzig and referred to in Chapter 8, is designed to harness expertise in the research community in all member-states to provide intellectual support and research advice for those responsible for preparing the *ESDP*. The initiative has been strongly promoted by the German government. Agreement in principle to proceed with the concept was taken at the Informal Council of Spatial Planning Ministers under the German presidency in Leipzig in September 1994, and the go-ahead given at the Strasbourg Informal Council in March 1995 in anticipation that progress on the *ESDP* would be such that the services of the network would be required later in 1995. After the March council, however, the momentum slowed on the *ESDP* project, reducing the pressure to make the network operational.

The exact form that such a network would take was not resolved at Strasbourg and will not be until 1996. The concept is that a network of research institutes from each member-state, selected for their expertise, will act collectively as an observatory of spatial trends and change, and of the impact and effectiveness of EU spatial policy, throughout the EU. It was felt that, if spatial policy was to be more precisely developed in its next phase, the advice and research capacity of such an observatory would be an essential underpinning.

There has been some discussion about whether the emphasis should be on a network of people with expertise on EU spatial policy issues in universities and research institutes throughout the EU, with a minimum of central co-ordination, or on a body of researchers in one place with national members of the network as advisers. The latter is favoured by those who see it as a European version of the German government's spatial research institute in Bonn, the BfLR, while the former is favoured by those member-states who do not wish to encourage the rapid institutionalisation of EU spatial policy, and by those who fear the financial implications.

The parallel network for spatial research institutes in the CEE, promoted and funded by the German government (see Chapter 8), is operational and functioning primarily as an information exchange and advice facility.

Compendium of Spatial Planning Systems

The *Compendium of Spatial Planning Systems and Policies*, to give its full title, will, when it is complete and available, provide the most comprehensive information yet about the planning systems and practice of member-states. A full discussion of the compendium would be more appropriate in a book on the comparative analysis of national planning systems, since one of the main purposes is to provide planning practitioners and anyone involved in the development industry with detailed information on the planning system and procedures in each member-state, and what the main directions of policy are.

It is appropriate to refer to it here since the project was commissioned at the same time that the *Europe 2000+* and the transnational studies were underway, and it is being prepared on behalf of the same division of DG XVI. Preparation of the compendium is being undertaken by consultants under a contract let in 1993 by DG XVI, and it includes all 15 member-states. A parallel study of Norway is also underway, funded separately by the Norwegian government.

The objectives are twofold: to identify key issues at the EU level; and to make available for anyone who needs to know (developers, professional firms, policy-makers, lobbyists and pressure groups, etc.) detailed information on the planning systems, regulations and policies and the operations in practice of the planning system in each of the 16 countries.

Work on Part I of the compendium was nearing completion by late 1995 (Shaw *et al.*, 1995). This is intended to focus on the spatial planning systems themselves, and will contain for each country included an overview; an account of the planning authorities and institutions at national, regional and local levels; instruments for achieving planning policy objectives; the mechanisms used to regulate and control development; and any special agencies or mechanisms designed to make implementation more efficient and effective and to conserve the natural and built environment. A series of national reports will be accompanied by a comparative analysis.

Part II will focus on spatial planning policies of the countries analysed, and includes a series of case studies. Each national study will include seven case studies of projects that have been completed or are under contruction, and which have completed all the planning procedures at the time of the study (Nadin *et al.*, 1995). These are designed to show how planning operates in practice in each member-state, and are also chosen with an eye on the interrelation of different levels of government, the implementation of EU initiatives and their integration into national policy-making structures. The list of case-study topics also gives an indication of the themes considered to be of greater EU interest.

Of the seven case studies, five will be on topics common to all countries and two will be selected from a list of possible themes, subject to the constraint that each topic chosen must be selected by at least three countries in order to allow comparative analysis. The full list of case-study topics is as follows:

- TENs.
- Regional policy and application of the Structural Funds.
- Crossborder co-operation, joint strategies and use of INTERREG.
- Management of urban growth, including new settlements and major urban extensions.
- Coastal planning, balancing pressures from tourism and recreation, conservation and industry.
- Environmental policy and impact assessment, and their integration with planning procedures.
- Industrial decline and revitalisation of abandoned areas.
- Neighbourhoods in crisis, to show integrated responses to regeneration of urban areas with social and physical problems.
- Rural revitalisation, rural economic development and tourism, and use of the LEADER programme.
- Heritage preservation and conservation of historic areas.
- Special events, to show the role of major sports or cultural activities as vehicles for urban or regional regeneration.

Any project of this sort is very soon confronted with the rapid rate at which planning legislation is introduced or changed. Hardly a year passes without new or revised legislation being enacted by at least one national parliament, and the rate of legislative change in planning is, if anything, increasing during the 1990s. The compendium is intended to be capable of being updated, but the problem of keeping up to date is never easy to solve.

Other sources

Section C of *Europe 2000+* summarises the main features of the planning systems of the then 12 member-states, but only one or two pages are devoted to each country. The compendium is able to build upon a number of more comprehensive studies of national systems of planning, although there has not been any previous study of comparable detail which has included all member-states.

The study commissioned by the UK government entitled *Planning Control in Western Europe* (Davies, 1989) covered in considerable detail the systems for the control and authorisation of development in Denmark, France, the former Federal Germany and The Netherlands, analysing these alongside that of England but not the other parts of the UK.

In 1991 the German government, concerned for the competitive position of Germany in the single market, commissioned a study of urban land and property markets in a selection of countries that were either key competitors or whose system and recent land policies displayed some features of comparative interest.

This project covered only five countries, France, Italy, The Netherlands, the UK and Germany. The analysis comparing these five countries has been published by the German government (Dransfeld and Voss, 1993), and a series of books based on the five national reports (plus a study of Sweden) has also been published (Acosta and Renard, 1993; Dieterich *et al.*, 1993; Needham *et al.*, 1993; Williams and Wood, 1994; Kalbro and Mattsson, 1995; Ave, 1996).

Although the compendium project was an initiative of the Regional Policy Directorate, the Environment Directorate has not totally relinquished its interest in planning and land use since the *Green Book*, and of course it is in the environment title of the treaty that 'town and country planning' and 'land use' are included, from which Figure 4.1 (p. 58) is taken.

The money for the compendium project, like that for *Europe 2000+*, comes from the Article 10 budget of DG XVI. Why are they spending money on this? A major reason is to provide better information for the professional planners, developers, consultants and other actors in the business.

Another motive must be to provide the Commission with the information on which to base proposals for EU legislation. A number of circumstances in which this may be considered have been identified in previous chapters, including the need to expedite TENs by co-ordinating planning procedures along the route, crossborder planning and INTERREG projects, and the possible introduction of an EU urban policy. One thing not on the agenda, because the complexities far outweigh any theoretical benefits as the Compendium will amply demonstrate, is any question of a harmonised local planning system.

European Planning Atlas

The concept of a *European Planning Atlas* was promoted by the German government in 1993, the year preceding their presidency. It is not going ahead in the form envisaged then, and did not form part of the package endorsed by the Leipzig Council, but the concept may be revived and is of interest as evidence of the German government's concern for EU spatial policy.

Their promotion of the idea is a reflection of their belief that the EU needs spatial policy concepts and guidelines, and recognition, from their perspective, that there is ever-growing demand for spatial policy coherence at the EU scale because both public and private investors are increasingly thinking in European terms.

The German government's ministry responsible for spatial planning, with the support of DG XVI, invited other governments and academic and professional experts to a conference in Bonn in 1993. The objective was to exchange views and debate the planning-atlas concept, and decide whether to adopt a resolution in support of pursuing the proposal.

In essence, the idea was that a network of contributors should prepare material which would set out, in an easily understood form employing innovative cartography, the different spatial policies and their territorial extent throughout the EU. It would cover the range of EU spatial policies, including for example the boundaries of all Objective 1, 2, 5(b) and, since 1995, Objective-6 areas, less favoured areas, areas eligible under Community Initiatives or other programmes. It would also include those main spatial policy designations arising from national and regional policies of member-states which are most likely to be of interest to, or offer useful initial guidance to, developers from other member-states. In this category, the areas designated as national or regional parks, nature protection areas, or which benefit from regional assistance or other broad spatial policies, could be designated.

The *European Planning Atlas* has been overtaken by the compendium project, a much larger project than the atlas proposal was ever likely to be. The compendium should meet many if not all the objectives sought by the atlas proposal. However, there may still be a place for an accessible source of cartographic reference to the spatial policies of EU significance, especially if the updating of the Compendium proves more cumbersome than its compilers hope.

Conclusion

The spatial problems that need to be addressed (accessibility, peripherality, disparities of quality of life and economic prosperity) are well known and substantiated in many reports such as the periodic reports published by DG XVI.

The central questions concerning the next stage of EU spatial policy-making are linked to issues that go to the heart of EU politics and the whole project of European integration, namely, the issues of the national sovereignty and the relationship between member-states and the EU institutions, and of the single currency. It is noteworthy that *Europe 2000+* includes a preliminary analysis of the territorial impact of public finance and of instruments for redistribution and equalisation. The significance of this as an instrument of spatial policy did not escape the rapporteur appointed by the Regional Policy Committee of the EP, whose report makes a point of welcoming it (EP, 1995: 7).

Any changes in the treaties, or new powers over any of the sectors of spatial policy reviewed in the preceding chapters, will be awaited with great interest or foreboding, as the case may be. Current political reality is recognised in the emphasis on subsidiarity and the non-binding nature of the *ESDP* in its political principles, but this inevitably means that the weakness identified for *Europe 2000+*, that it lacks legal authority, will be perpetuated unless changes are agreed at the IGC. These issues are the subject of Chapter 15.

PART III

Operating in Europe

13

The professions and planning practice

Many people want to take advantage of the opportunities created by European union to seek employment or extend their professional practice beyond the confines of their own country. The right to do so, in the form of the right of service and the right of establishment, was enshrined in the Treaty of Rome (see below) and has been reinforced in the SEA and Maastricht treaties. The opportunity to do so depends of course on securing employment by demonstrating qualifications and expertise that convince prospective clients or employers.

For spatial planning professionals, the situation is especially complex. They are confronted not only by different concepts of what constitutes a profession and how it is defined but also by different concepts of planning and the skills expected of planners, and above all by the fact that the profession of spatial planner, however entitled, is not recognised as existing in all member-states.

The function of spatial planning exists and is practised in all member-states, but the academic qualifications and professional title, if any, of those who carry it out vary greatly. The *Compendium of Spatial Planning Systems* (see Chapter 12) will, when it is available, offer a much more detailed basis on which to prepare for practice in another member-state than has hitherto been available. Planning is not alone in facing great disparities in status and recognition in the EU. For different reasons two professions closely associated with planning, architecture and surveying, also do. These are also considered in this chapter.

Broadly, and at risk of drastic generalisation, professions can be grouped into three categories: those which are recognised as a profession in each member-state and about which there is broad consensus concerning the nature of their expertise and of the purpose and scope of the service they offer (e.g. dentistry); those which exist in all member-states but for which the same degree of consensus concerning the scope and range of their responsibilities is lacking (e.g. architecture); and those which are not acknowledged at all in some member-states, while being well established in others (e.g. town planning).

Professional titles

There are other dimensions which cut across any attempt to take a pan-European view of professions. One is the question of whether they are subject to any legal definition or legally based protection of professional title. This is normally the case for the medical professions, but the incidence of legislation varies greatly among the professions concerned with spatial policy. The professional title of 'architect' is subject to legislation in several countries including the UK under the Architects Registration Acts of 1931, 1938 and 1969.

In some cases, the architecture profession is so defined that it includes expertise in town planning and, in the extreme cases (such as Spain), anyone who does not hold the title architect is limited in the extent to which he or she can undertake professional work in spatial planning. The other end of this spectrum is represented by the UK, where the town planning profession has not merely existed but been recognised by government and the universities at least since the 1940s (Schuster, 1950).

Linked to the issue of legal definition is the question of accreditation. Professional groupings or associations have in most countries come into existence and advocated for some years that planning should receive recognition. In most countries, achievement of this objective is marked by legislation defining the profession and professional title, and specifying the qualifications required in order to use that title, plus any other regulatory measures deemed necessary to protect the public. The question of accreditation in these countries is thus a matter for legal regulation.

The main exceptions to this principle are found in the UK and Ireland, where a system of regulation and accreditation governed by non-government professional associations is normal in the spatial planning field. Entry into the profession, and regulation of professional title, is undertaken by an institute recognised for this purpose by, but independent of, the national government.

In the UK, for example, the first university to offer a course in town planning was the University of Liverpool in 1909. The RTPI was founded in 1914, but general acknowledgement of the role of the profession was not attained until the time of the Town and Country Planning Act 1947 and the establishment of the first undergraduate degree in town and country planning in Newcastle (then Kings College in the University of Durham) in 1946. The RTPI received its royal charter, effectively the final stage of the recognition process for the profession itself, in 1971.

The cultural dimension, in respect of the concept of a liberal profession, is another factor distinguishing different professional traditions. In the traditional French way of thinking, a liberal profession is composed of people who offer their clients their best advice, based on their professional expertise and experience but totally independent of any government employer. In other words, members of a liberal profession are consultants paid by clients' fees. The alternative concept is that a person can possess professional experience and expertise and be employed by a government department or large corporation, who require such advice on a continuing basis.

The concept of the public sector professional has a long history in Britain, especially in local government, but it is well established in several other countries. The British town planning profession has evolved in this context, and it is clearly understood that members of the RTPI have the professional duty to advise their political authorities on the basis of their best professional judgement, not on the basis of political instructions. This conflicts with the corollary of the liberal profession concept, namely, that a public official is a *fonctionaire*, advising and acting always under the direction of political authorities.

The whole subject of professional titles is most emphatically a theme where the terminology used can create misunderstandings and where the dictionary translations can often hide real differences in concept or exaggerate others, as the discussion of town planner, *urbaniste, planologen, Städtebaukundigen*, etc., in the context of the ECTP illustrates (see below).

Freedom of movement

Freedom of movement within the EU was set out as an objective in the Treaty of Rome. Article 52 states that 'restrictions on the freedom of establishment of nationals of a member-state in the territory of another member-state shall be abolished by progressive stages', and also that 'freedom of establishment shall include the right to take up and pursue activities as self-employed persons, and to set up and manage undertakings'. These two provisions, known colloquially as the right to establishment and the right to service, were subsumed by the four freedoms of the SEM (see Chapter 3).

The Commission has attempted over the years to put these principles into practice by putting in place agreed systems of mutual recognition. Initially, the approach was sector by sector or profession by profession. Directives were adopted covering several of the medical and paramedical professions, plus after many years of negotiation, the architecture profession in 1985. By 1986, eight such directives had been adopted (Owen and Dynes, 1993: 119–20).

Negotiations for the architects' directive took 17 years, giving it the dubious distinction of being the product of one of the longest periods of negotiation on record. This experience, and recognition of the great disparities in concepts of the professional and academic prerequisites for practice in professions such as architecture that reflect national cultural differences, led the Commission to adopt a new approach.

In December 1988, the Social Affairs Council adopted a directive 'on a general system for the recognition of higher-education diplomas awarded on completion of professional education and training of at least three years duration' (Directive EEC/89/48, *OJ* L 19 24.1.89). The objective was to require member-states to create a framework within which a whole range of professional qualifications could be mutually recognised so that individual citizens holding any of these qualifications could have their qualifications recognised and register for professional practice in another member-state within a reasonably short period of time.

The directive sets out broad definitions of its terms in order to encompass the wide range of practices noted above. A regulated profession is one whose activities or membership are subject directly or indirectly to legislation or the authority of non-government associations recognised by government for this purpose. Diploma is short-hand for any degree, diploma or other professional qualification awarded by a university or other competent authority, including a professional institute, authorised for this purpose in a member-state.

The basic principle adopted is that member-states cannot refuse registration on the same conditions that apply to its own nationals if the applicant holds qualifications from another member-state or has practised the profession in question full time for at least two out of the previous ten years in another member-state which does not regulate that profession and has appropriate higher-education diplomas. However, the applicant may be required to undergo either an adaptation period or an aptitude test before registration.

An adaptation period is a period of supervised practice under the direction of a qualified person, plus possibly some further training and an assessment. This period cannot be required to exceed three years.

An aptitude test is an examination of the professional knowledge and skills of the applicant, concentrating on those aspects, such as national planning law, which are essential to practice but which were not covered by previous education. The aptitude test cannot repeat the basic professional examination already passed in the applicant's own country, but it can include a test of language ability. Language competence remains in many respects the greatest single impediment to free movement of professionals.

This directive came into force on 4 January 1991. Annexed to it is a list of professional associations to whose activities it applies. In the spatial planning field, the English text lists the Irish Planning Institute and the RTPI, plus the Royal Institution of Chartered Surveyors among five Irish and 38 British associations. Other language versions list bodies in the countries concerned. The original list was not intended to be final, and other associations could be added. One that has since been added to the list is the British Landscape Institute.

The European Council of Town Planners

The European Council of Town Planners (ECTP) or Conseil Européen des Urbanistes, is a federation whose members are the professional town planning associations and institutes from EU member-states. It provides a focal point for the profession in the EU and a forum for the exchange of views and information, especially concerning planning practice as it is affected by single-market and EU legislation and Commission policy initiatives. Its members agreed to apply common regulations on planning practice, education and training.

Its fundamental purpose is to defend the professional title and practice of town planning as a profession within the EU. The aims of the ECTP are

1. To promote the freedom of professional town planners to practise throughout the member countries of the European Community [*sic*] by the mutual recognition of their professional qualifications and competence.
2. To represent and promote the profession of town planning by defining the scope and nature of its activities; to ensure recognition of its values and foundation in the policy of land management, taking account of European institutional and governmental systems, social and economic processes and community aspirations.

(ECTP, 1993)

A	No member
B	Belgische Vereniging van Stedebouwkundigen (BVS); Chambre des Urbanistes de Belgique (CUB); Vlaamse federatie voor Planologie (VFB)
D	Vereinigung für Stadt- Regional- und Landesplanung (SRL)
DK	Foreningen af Byplanlaeggere (FAB)
E	Asociación Española de Técnicos Urbanistas (AETU)
F	Société Française des Urbanistes (SFU)
GR	Greek Planners Association (GPA)
I	Associazione Nazionale degli Urbanisti (ANU)
IRL	Irish Planning Institute (IPI)
L	No member
NL	Bond van Nederlandse Planologen (BNP); Bond van Nederlandse Stedebouwkundigen (BNS)
P	Sociedade Portuguesa de Urbanistas (SPU)
S	No member
SF	No member
UK	RTPI

Figure 13.1 *Members of the ECTP*

Full membership of the ECTP is open to professional bodies from all EU member-states, and can include more than one from a member-state. Bodies from non-EU countries may become corresponding members, and associate membership is open to bodies for whom town planning is a secondary but not primary purpose. There is no individual membership. The ECTP membership is listed in Figure 13.1.

The ECTP was founded in 1985 and is established under Belgian law as an international association with a learned purpose, with its registered office (*siège sociale*) in Brussels. It is designated as a Liaison Committee with SEPLIS (Société Européenne des Professions Libérales Indépendantes et Sociales). Its official languages are English and French and it meets as a formal council twice each year.

The ECTP undertakes lobbying and promotional activities in pursuit of its aims, fostering contacts between members of its constituent bodies, defining and promoting the concept of town planning and its educational requirements, and commenting on proposed EU policies and directives. The main links with the Commission are with DG III (Internal Market), DG VII (Transport), DG XI (Environment) and DG XVI (Regional Policy). On behalf of the latter it runs the scheme for European Urban and Regional Planning Awards.

Origins of the ECTP

The origin of the ECTP lies in a 'Liaison Committee of National Institutes and Associations of Professional Town Planners within the European Economic

Community' which was created in the early 1980s by the RTPI, the then Belgian associations, the French SFU and the Dutch BNS.

The Liaison Committee adopted an International Agreement and Declaration in November 1985 which paved the way for mutual recognition by agreeing to formulate a common definition of the field and nature of the town planning profession, the education and training criteria and professional conduct requirements. One of the clauses in the declaration states the problems it was facing:

> There are no existing legal obstacles to the free movement and right of extablishment of town planners within Member States of the Community but there are substantial differences as between the various Member States in the definition, purpose, role, scope, structure and implementation of town planning; in the organisation and structure of the town planning profession; and in the training, competence and codes of conduct of professional town planners.
>
> (Liaison Committee, 1985)

Much progress has been made since by the ECTP in defining planning and in helping members to learn from each other, but many of these problems remain as non-tariff barriers. The Compendium (see Chapter 12) should provide the basis for overcoming them further.

Several of the members of ECTP have very small memberships, and membership of the British RTPI outnumbers all the rest put together. In some member-states, these associations face a struggle for recognition either because planning is not recognised as a separate profession and another more powerful profession such as architecture still controls it, or it is undertaken by people educated in university departments of economics or geography that fall outside its scope. The process familiar in the UK and Ireland of professional accreditation of degree programmes by a non-government professional body is almost unknown elsewhere. The origin as a Liaison Committee reflected in part attempts by some of its smaller members to gain recognition by associating themselves with the RTPI.

Title of the profession

The profile of the ECTP in some countries is limited by the incorporation of the term 'town planners' and *urbanistes* into its official titles in English and French. 'Town planning' normally translates into German as *Städtebau*, not *Raumplanung*, and in French it is the equivalent of *urbanisme* not *aménagement du territoire*. The latter is the accurate translation of *Raumplanung* or *Raumordnung* or 'spatial planning', and is closer in sense to the professional activity with which many planners who do not also hold architectural qualifications would identify themselves.

The ECTP is sometimes seen therefore as representing a particular view of planning and of professional practice at the local urban-design and project-development scale. The wider spatial policy concerns of the *Raumplaner* or *aménageur* are not demonstrated by use of the term 'town planner' in non-English speaking countries although it is the accepted generic term in the UK and Ireland. For this reason, its membership in several countries falls far short of representing all who are professionally concerned with spatial policy in the sense used in this book.

The Dutch responded to this problem by setting up a new association, the BNP, to represent the *planologen* who have a public-policy based education rather than a technical education as the members of the BNS normally have. The VFP was subsequently set up for similar reasons by the Flemish planners in Belgium.

Related professions

In order to provide a perspective on the situation of the planning profession in the EU, it is helpful to consider two professions with overlapping interests, the architecture and surveying or real-estate professions.

Architecture is recognised as a profession in all member-states, in several of which planning is claimed by the architects to be a part of their expertise. Separate planning institutions and university degree courses find it difficult to assert their independence and gain recognition for planning as a field of study and profession in its own right.

Surveying and real estate, on the other hand, are not necessarily recognised as professional activities at all, or if they are, then it is as more than one activitiy. It therefore represents a case of competition between systems within the SEM, as well as competition between firms.

Architecture

Architecture is the only profession close to the area of expertise of planners that is the subject of a directive on mutual recognition of qualifications, the right of establishment and the right of service. This directive, which was adopted in June 1985, is one of the longer and more complex pieces of EU legislation in the field relating to spatial planning, running to 32 articles (the environmental assessment directive, discussed in Chapter 10, by contrast, has only 14 articles yet is much more wide ranging in its significance). The experience of preparing this directive, and the time taken to achieve adoption, was one of the reasons why the Commission decided that it was necessary to find an alternative to the sectoral profession-by-profession approach – if mutual recognition was to be achieved for a large number of professions on a reasonable timescale.

The architecture directive, Directive EEC/85/384 (*OJ* L 223, 21.8.85), goes into some detail about the contents of courses of study. This includes, in addition to all other aspects of architecture, specification of 'an adequate knowledge of urban design, planning and the skills involved in the planning process' (Article 3.4), 'an understanding of the relationship between people and buildings, and between buildings and their environment' (Article 3.5) and 'an adequate knowledge of the industries, organisations, regulations and procedures involved in . . . integrating plans into overall planning' (Article 3.11).

The directive specifies the minimum length of the academic component of an architecture course as being the equivalent of four years' full-time study, although it provides for graduates holding the three-year diploma from German *Fachhoch-*

schulen to be accepted with sufficient experience. This was one of the difficulties holding up negotiations on the directive, since four years is actually shorter than the minimum periods of study normally required in most member-states.

The directive lists the degrees and diplomas in architecture, and the awarding universities and institutions, from each member-state. The system of accreditation that operates in the UK allows the directive to specify that all courses recognised by the Royal Institute of British Architects and the Architects Registration Council of the UK are accepted. For all other member-states, all individual diplomas are listed, so the list could soon become out of date.

The directive was accompanied by a decision (Decision EEC/85/385, *OJ* L 223) setting up an Advisory Committee on Education and Training in the Field of Architecture. This committee has the duty to promote a comparably high standard of architectural education throughout the EU by exchanging information, developing common approaches to the structure and content of courses, and ensuring that curricula respond to new social, scientific and technical developments and to environmental protection. The committee consists of three experts from each member-state, one practitioner, one academic and one from the competent authority for accreditation, such as a professional body or regulatory authority.

Since the directive requires that each member-state should recognise all these as a licence to practise on the same terms as its own qualifications, there was some concern on the part of institutions representing qualified planners who did not hold architecture qualifications that the existence of this directive might inhibit their right to practise in other member-states. For this reason, the Liaison Committee and the ECTP, when they were founded in 1985, were initially lobbying for a directive on planning similar to that for architecture, so that formal recognition of the professional title of planner, and the right to practise under that title, was established for all member-states.

Chartered surveyors and the real-estate profession

The UK profession of chartered surveyors, whose profession is regulated and title protected by the Royal Institution of Chartered Surveyors (RICS), faces far greater disparities in the EU context than the planning profession does. Although most of the professional services offered by its members are called for and offered by someone in other member-states, the particular range of skills and expertise represented by the professional title 'chartered surveyor' are not paralleled in any other member-state except Ireland.

Even the terminology is very specifically British. The title used for equivalent professional groupings in other languages varies greatly, often because different aspects of surveying expertise, such as valuation, land surveying, property investment advice and project development, are undertaken by holders of different professional qualifications in different countries. The American term 'real estate' comes closer than any British term to the sense of the terms used in most other major European languages to define the central expertise of a property profession (van Breugel *et al.*, 1993). For this reason, the association that was founded in 1994 to bring property researchers and educators together with practitioners across Europe, the European Real Estate Society (ERES), chose this term for its title.

The scale of the European real-estate industry is enormous, as the success of the international property fairs held each year in Cannes under the title MIPIM demonstrates. MIPIM started in 1990 and now regularly attracts over 6,000 participants representing over 2,000 companies, including firms of developers as well as professional advisers, from around 50 countries. The organisation behind this is a private commercial enterprise which had identified a gap in the existing pattern of conferences and conventions. MIPIM is essentially an enormous market-place at which all the leading surveying and other property development firms in all sectors of the land and property market must be present if they are to remain competitive.

Many firms of British origin have established themselves in major cities of most EU countries where they are competing with firms which come out of very different concepts of professionalism or ways of doing business in property. This process has been underway since the 1970s, and many of the big names of the British profession have offices in cities in several EU countries. Their signboards are a familiar sight in Brussels, for example.

The option of forming a federation with equivalent professional institutes in other member-states, which is available to most professional bodies, was not open to the RICS even on the unequal partnership basis characteristic of the ECTP. The RICS therefore decided upon the alternative of setting out to create a European profession of chartered surveyors, the European Association of Chartered Surveyors (EACS). This has affiliated associations in most member-states. The RICS has also embarked on the road of accrediting university degrees and diplomas offered outside the UK and taught in other languages. The first course to be so accredited as meeting academic requirements for RICS membership is a postgraduate diploma in real-estate management at the Immobilien Akademie of the European Business School at Oestrich-Winkel near Wiesbaden, Germany.

Another feature of the European strategy of the RICS was the opening of an office in Brussels in 1993. In many ways, this can be seen as part of the lobby in Brussels and has similar purposes. Its four main objectives are to

1. monitor EU legislation which is likely to affect the market in which chartered surveyors operate;
2. lobby the European Commission, members of the EP and other Community institutions on behalf of the profession and the industries it serves;
3. promote the chartered surveying qualification as the *premier* qualification for the property profession throughout Europe; and
4. help keep key property issues on the agenda for European decision-makers.

Professional education

The variety of educational structures and concepts of what constitutes the appropriate disciplinary foundation for professional education in spatial planning, the extent to which degree programmes in spatial planning are available at all, and the esteem with which they are held, vary greatly in Europe. In many ways, this variety matches the extent to which planning is recognised as a distinct professional activity and the different concepts of the nature and skills of the profession. It is difficult to assign cause and effect since educational structures have developed in the professional context of each country, and vice versa.

In some countries, planning education is closely associated with the traditions of technical and design-based courses in architecture and civil engineering; in others it is offered within a social science or policy science framework. Another dimension is whether it is available as an undergraduate or postgraduate programme, and whether it is seen as a discipline in its own right or whether planning is taught as a component or supplement to degrees based on some other central discipline, such as architecture or geography.

From the point of view of the spatial policy-maker at the EU scale, and of the individual professional who wants to practise in another country, these differences and the cultures of planning they represent and support form an additional element in the matrix of non-tariff barriers which must be surmounted.

The Association of European Schools of Planning (AESOP; see Chapter 8) has, since its foundation in 1987, played a major role in bringing together planning educators from universities throughout Europe who represent these different educational and professional traditions. It enables them to gain a vastly improved understanding of these differences.

AESOP has made it easier for university planning schools to get together to participate in the ERASMUS programme. ERASMUS (European Community Action Scheme for the Mobility of University Students) is a programme that has been in operation since 1989 to encourage university departments teaching the same subject in different countries to form networks within which staff and student exchanges can take place. The ambition is that up to 10% of all EU students will have the opportunity for some element of study in another country. Planning education is well represented in ERASMUS and has stimulated substantial sharing of experience. ERASMUS is due to be replaced by the SOCRATES programme in 1996 (see Appendix II).

AESOP does not itself have any standing as a professional body, but the enhanced understanding of the different national traditions in planning education that it has acquired is a valuable resource on which the ECTP could draw. Although written before AESOP and ERASMUS were fully established, the study by Rodriguez-Bachiller (1988) remains the best overall analysis of the different forms of planning education in Europe and their underlying concepts, drawing comparisons with North American practice. If a similarly comprehensive study of European planning education were to be done in the mid-1990s, it is probable that some convergent trends would be evident.

For the cognate professions, architecture education is supported by its European association, the European Architectural Education Association (EAAE), whereas the European Real Estate Society and the European Foundation for Landscape Architecture both include their respective educational and academic sectors alongside professional practitioner membership.

International planning associations

There are a number of international planning associations which are not specifically European in their focus but which do have strong European membership and participation in their activities. Examples include the International Society of City and Regional Planners (ISoCaRP), the International Federation of Housing and Planning (IFHP) and the International New Towns Association (INTA). These all provide their membership with opportunities to gain a better understanding of planning practice in other European countries and to gain ideas from projects and concepts which may be worth applying in their home countries. However, none of them have the comprehensive reach which the ECTP and AESOP have in their respective spheres of interest.

Conclusion

In the report it commissioned in 1994 (RTPI, 1994), the RTPI sets out scenarios for its involvement within the profession as it evolves in the EU. The architecture profession has no alternative but to work with equivalent professions within the EU context – the RTPI's responsive scenario. The real-estate profession, or more specifically the British chartered surveyors, have decided that they have no alternative but to adopt a very proactive policy of taking the lead in setting up an EU professional association, the EACS, and encouraging educational developments in other member-states that follow their own accreditation guidelines – the RTPI's proactive scenario. It remains to be seen how influential the ECTP is likely to become, how far the interests of national professional bodies are served by it, and how far the freedom of movement and recognition of title of the individual professional planner is protected by its intervention.

14

The European functions of local planning authorities

Many local and regional authorities have staff appointed specifically to handle EU affairs. In some cases, a whole section or group of staff is employed for this purpose; at the other end of the spectrum this function may be only one among several duties of one person. It is not unusual for spatial planners to be appointed to such posts, or for planning departments to assign the task of EU liaison to a member of the professional staff of the department, especially if it is also responsible for economic development. As a result of the substantial crossborder element in its work with Nord-Pas de Calais and the Euroregion (see Chapter 8), Kent County Planning Department was the first to assign a professional planner full time to European planning work. For many smaller authorities, another way in which EU affairs may be handled and the necessary expertise obtained is through the appointment of consultants.

The expertise required for such tasks broadly corresponds to the subject-matter of Parts I and II of this book. In other words, such people are expected to have a good understanding of the EU policy and decision-making processes, and of all the policy sectors, programmes and legislation relevant to the situation of their employer. Their prime duties often lie in securing money for projects within their authority by preparing applications to the Structural Funds or bidding for funding from Community Initiatives or other programmes. Duties may also be in support of the authority's networking and partnership activities. Sometimes, promotion of the name and image of a city within the EU is more important than direct funding, and sometimes the objective is to become an influential voice within EU associations of local or regional authorities and networking and lobbying organisations. The emphasis varies depending on whether the area comes wholly or partly within an objective area, as well as on the political objectives and strategy of the authority.

In general, in addition to the knowledge base, the necessary skills include the ability to think strategically, spot opportunities and take initiatives, and negotiate within the authority, with other government bodies and the Commission, and with representatives of authorities in other countries. Collaborative and reticulist skills, which planners often possess, are of the essence, as are good advocacy skills. Language skills are also often expected or assumed, especially in English and French.

Arrangements vary greatly in different authorities and in different member-states. The contrast between the UK and Germany is quite striking in respect of the European liaison function of municipalities, and is discussed below. Most other member-states fall in between these two situations, but it is only possible to give some general principles and indicate the range of functions performed by people holding these appointments.

Patterns and principles

In general, in member-states with a well developed regional or federal system of government, European liaison functions are left to the state or regional authorities, whereas in countries without a well developed regional level of government, municipalities are more likely to employ their own European officers. Thus, most large British cities, especially in Objective-2 areas, have had someone designated as a European liaison officer for some years, while in Germany this has not been seen as a necessary function for a municipality.

This is demonstrated by the case of Duisburg in the western part of the Ruhrgebiet on the Rhine. The city of Duisburg is located in an Objective-2 area and is as much in need of industrial revitalisation as many British industrial cities in Objective-2 areas, yet it has only appointed its first European officer in the mid-1990s and is the first German city so to do. In fact, such is the disparity between German and British participation rates by municipalities in networking programmes such as those funded by RECITE (see Chapter 8) that the RECITE office in Brussels has conducted an investigation into the reasons for this situation. Some of the findings are discussed below.

Another key variable is the extent to which municipalities are in competition with others from the same region or member-state for the funding opportunities that are on offer and need to be bid for. In several Objective-1 areas, either the respective central government departments take charge of bidding for the Structural Funds, or the amount of funding available is such that there is little competition to be faced by a local authority in obtaining funding for its own projects. In this situation, there is little call for a European affairs official to take a proactive and promotional role. The function then becomes more one of administration and accountancy, and it is the responsibility of the normal planning office to authorise projects that are funded and to integrate them within the spatial plans of the authority.

On the other hand, in situations where there is no shortage of suitable projects that would qualify to be supported by the Structural Funds, cities may face competition to get their project included within submissions from their national government to the Commission. In the first place, such competition is within their own country. However, a European officer may be able to gather evidence from cities in other member-states which would support the case by showing, for example, that a particular project fulfils both letter and spirit of the Commission's guidelines better than others, and so is a safer prospect for

Commission approval. European officers may also find alternative funding opportunities under less well-known Commission programmes or identify precedents in other member-states or interpretations of the guidelines that are followed elsewhere and which may be deployed to advantage.

A third variable that may influence whether a local or regional authority decides that it should appoint a European officer is whether it believes that its case is well understood and backed by its own national government, and that its national government will advocate effectively with the Commission whenever necessary. Sometimes, the bypassing-the-capital effect, noted in Chapter 8, is a motivation for such an appointment.

Function and role

The core purpose commonly is to secure maximum potential financial benefit from whatever objective status the area has or from any other EU policy that might be applicable to the category of authority, and the preparation of strategies and the influencing of spatial planning policies in order to ensure that the authority takes advantage of any opportunities for funding that are available. It is not always a question of currently available programmes. Ideally, an EU liaison officer must try to make sure the authority is ready to go for whatever new opportunities may come along following reviews of existing policies or changing eligibility criteria.

A second purpose is to ensure that the authority participates fully and effectively in the appropriate European organisations and develops partnerships with authorities in other member-states wherever such association would be beneficial. One value of this is the promotion of causes of common interest, and presenting the case for support for these interests in Brussels and to national politicians.

Development of partnerships and identification of networking opportunities that are worth while are important. It is not wise to join everything available but it is desirable to be able to advise on which ones are worth joining and what influence or opportunities may be obtained thereby. For example, when changes in eligibility criteria for Community Initiatives are proposed, some organisations will be consulted at length by the Commission, and authorities which are influential participants in that organisation may find that changes are beneficial to them.

The EU liaison officer often has a wide range of working groups and committees to attend within the authority in order to support and advise the many partnerships that are often required under a CSF or other programmes. These may be with development agencies' educational and training institutions, cultural and trade associations and the social partners, employers and trades unions. The supply of information about EU programmes and new legislation or proposals to all sections of the authority is usually also an important internal function.

Externally, representation is often required on bodies such as the Assembly of European Regions or the Council of European Municipalities and Regions and on committees of any European lobbying and networking associations that the authority has joined. In fact, it is rarely worth joining bodies such as Eurocities or RETI (see Chapter 8) unless sufficient staff time and resources are devoted to allowing full participation in meetings and working groups.

Other tasks may include searching for and evaluating possible partners, advice and support for other international activities (such as participation in environmental or energy programmes) and support for Council of Europe projects. Liaising with any Brussels office that the authority maintains or shares is also an important part of the job.

There can be an element of rivalry between neighbouring authorities and even between those from the same country, so partnership with municipalities or regions in other member-states can be doubly beneficial. As they are not rivals within the same programming area, they are able to present a more effective lobby because more than one member-state is represented, and each may gain from the stimulus of learning about the policies and planning ideas of the other.

Funding and grant applications

An important part of the job is the preparation of funding submissions and grant applications. The first stage is identifying the opportunity and securing agreement to proceed with a project that may be eligible for EU funding. This may involve developing ideas and plans on the basis of the categories of project for which a particular fund is designed. More often, it involves liaison within the authority to ensure that projects already proposed under existing policies, for example as part of a local economic development programme, are so presented or specified that they meet the criteria for funding.

The details of a project, the purpose which receives most emphasis, or its contribution to the economy of the surrounding area and complementarity with another project as part of an overall strategy, may need to be amended in order to maximise the chances of a funding application being successful. An understanding of the logic of EU policies is necessary in order to present a project to the best possible advantage from the point of view of the Commission.

The hard work comes in the next stage, that of completing the funding application forms. Much of the administration for the Structural Funds and financial instruments is handled within member-states by national or regional government, so the actual application forms vary. The government departments responsible for each fund in each member-state will all have their own application forms, although much of the information requested will be common to all in any specific programme.

The application forms

In order to demonstrate the type of information requested and the degree of detail, the scope of the grant application form issued by the UK Department of the Environment for ERDF applications for projects in England is outlined below. Similar forms are issued by the Scottish and Welsh Offices for the other parts of the UK, and by the Department of Education and Employment for the ESF and the Ministry of Agriculture, Fisheries and Food for the EAGGF and FIFG. The ministries responsible in other member-states will also have forms which cover much of the same information.

The ERDF application form is 17 pages long. Twelve pages ask for information under 11 headings, four contain notes on how to complete each section, and one gives the list of contact points and addresses of the Government Offices of the Regions, to whom the completed forms are sent.

Details of the applicant authority with contact person, title and location of the project come first. The latter must include reference to the SPD or Community Initiative and objective within which it fits. Then the type of project must be indicated in order to show how it fits within the objectives of the SPD or Community Initiative. There are 33 categories, and a project may have elements of more than one, in which case the percentage of total project costs represented by each category must be shown.

The section requiring the project description occupies three pages and includes financial, technical and implementation details. These include environmental information on location within a protected area such as a national park or environmentally sensitive area; any positive or adverse environmental effects the project may have on the biosphere, air or water quality, urban environment or the coast, together with measures taken to overcome these effects; and whether planning permission has been obtained and any environmental assessment undertaken. If the latter is required under Directive EEC/85/337 (see Chapter 10), it is illegal to give ERDF funding unless it has been done.

Financial information requested in this section includes details of any other EU funding for linked projects, or other EU funding from ESF or EAGGF for the subject of this application. More financial detail is required later on the form, where all public monies, whether grants or loans from local, national or EU sources, must be set out and the programmed expenditure indicated for five years.

The project objectives form another section, where a description must be given of how the project fulfils the objectives and the specified priorities of the SPD or Community Initiative. This is done both in the applicant's own words and by indicating which of a list of ten possible priorities are the primary and secondary targets. This section also includes a list of over 60 possible monitoring and evaluation criteria, grouped according to different categories of output: business and industrial, training, social, tourism, environmental, private investment, transport infrastucture and jobs retained or created. For those relevant to the project, the proposed achievement and timescale in months that applies has to be entered. Responsibility for monitoring rests with the applicant, who must also provide a clear description of the methodology proposed.

The next sections require details of the timetable for the project, both in respect of the period from land acquisition to physical completion and in respect of the contract and financial settlement. All sources of loans and grants from any public sector source, plus any private finance, must be set out together with a written justification of the level of ERDF grant requested.

A separate section is devoted to additionality (see Chapter 7), which seeks information on whether the project could go ahead more slowly, on a reduced scale or not at all without the ERDF grant. If any contract for works associated with the project exceeds the threshold, it must be advertised in the *OJ* and a procurement questionnaire completed. Details and a copy of the entry in the *OJ*, or an explanation of why it has not been advertised, must be enclosed.

Finally, the Commission is keen to extract the maximum publicity value from projects it funds, especially in countries such as the UK where public attitudes to the EU are often negative. An explanation of what publicity is to be given to the contribution of the EU to the project must be provided through on-site notices, permanent plaques and coverage in the local press or specialist journals.

Lobbying and networking

The lobbying and networking operations of local and regional authorities, discussed in Chapter 8, have become such big business that an organisation has been set up to hold regular conventions for directors and senior representatives of such authorities. Under the title Directoria, a series of three-day conventions has been held in Brussels since 1993 primarily for the purpose of enabling directors of authorities from throughout Europe to meet each other.

It is attended by around 1,500 people representing over 500 authorities. The main functions of Directoria have been to enable delegates to meet each other and discuss possible co-operation; the setting up of partnerships and networks; and the making of preparations for bids to EU and other funding programmes for which such crossnational partnerships are necessary. Directoria meetings also provide the opportunity to meet Commission officials, and include workshops and presentations at which exchanges of experience and advice on new programmes are offered.

From 1995 the Directoria will take on a new role on behalf of the Commission as the forum in which the Commission informs local and regional authorities of forthcoming calls for proposals under Article 10 of the ERDF regulation. Directoria cover the wide range of policy sectors, including economic rural and urban development strategies, spatial planning, infrastructure and environment policies. A particular emphasis is inter-regional and cross-border co-operation, enlargement and the opportunities to develop links to CEE countries making use of the PHARE programme. The most important function for many authorities, given the sense of competition between the cities and regions of Europe, is the opportunity Directoria offer to publicise their achievements and undertake some place-marketing.

Opportunities for local and regional authorities to bid for funding will continue to be available from time to time. During the period 1995–8 there will be opportunities for partnerships from groups of authorities to bid for funding under the Internal and External Inter-regional Co-operation programmes. In each case, authorities from at least three countries are required, one of which must be from an Objective-1 region. In the external programme, one must be from a non-EU country. There will also be further opportunities to propose Urban Pilot Projects, and a new spatial planning initiative (Commission, 1995d).

Political liaison

In addition to advising the elected members of a council or authority on EU affairs as they affect the authority itself, a European officer may be called upon to advise in the context of the activities of a European organisation. If the authority participates in some programme of the Council of Europe, or one of its politicians is nominated or elected as a representative on a Council of Europe or EU body, the officer or EU section will be called upon to support and advise that person even if he or she is not representing that authority as such, but is representing a wider constituency such as an association or regional grouping of municipalities.

From time to time, any authority may find one of its members nominated to the CoR, and both politicians and senior officials could be nominated to membership of Ecosoc. Such an appointment would generate substantially more advisory and administrative work for that authority although the nominee is representing subnational government or an economic interest in general. The authority concerned would normally regard the extra workload as worth while in return for the greatly increased information and awareness of EU initiatives, direct insight into decision-making and opportunity to influence policy.

Links with local members of the EP can be another aspect of political liaison that often is of mutual benefit and valued as much by the MEP as by the authority. The MEP needs to maintain close contact with the local community, local politicians and local issues, and can often advise a local or regional authority of opportunities and useful contacts, act as an advocate for that authority or locality or meet Commission officials or ask questions in the EP on its behalf. The task of the European officer would be to ensure that this relationship with an MEP is effective and mutually beneficial.

Influencing EU policies can be part of the job, not normally alone but as part of wider lobby. Seeking partners and participation in city networking would be partly for this reason, so that any case that is argued is put forward by as broad based a coalition as possible representing several member-states. For some purposes, a member of a network from an Objective-1 area or a Cohesion Fund country is necessary, either to meet requirements of the formal rules or to meet unofficial political expectations. Networking and partner search is often undertaken partly with this in mind, seeking suitable partners who can be expected to share the same view of what policy changes should be advocated.

Networking and liaison in relation to EU programmes and lobbying may also take place through associations of local or regional authorities, leagues of municipalities or associations of cities. These exist in all member-states, in some cases as a single national association, while in others there is more than one representing different categories of local and regional authority, different parts of the country or political groupings.

Place and contribution of EU officers

The City of Birmingham is an example of a city that devoted considerable resources to the EU liaison function. It has set up a European Task Force directly responsible to the Chief Executive. This chain of command, which will also be adopted by the Highlands Council to ensure maximum benefit from its Objective-1 status when it takes over in April 1996, is an indication of the central place that the European liaison function has in some LRAs for whom this is a priority. Several other authorities that have decided to set up a European office for the first time have also placed it in direct line of responsibility to the head of the paid service.

In some LRAs, the European liaison function comes within planning or economic development departments. Where this is not so, for example where the officer reports directly to the Chief Executive, close working relationships with the spatial planning department are desirable. These working relationships are normal in order to ensure that the economic development strategy is integrated with the spatial plans that exist at local and regional levels. They also ensure that any projects proposed for funding fit within this strategy, have been subject to any necessary environmental assessment, and will receive authorisation. In the case of Birmingham, for example, the objective is that 'both work side-by-side to ensure that a consistent image is presented to the Commission (and to the UK government) of Birmingham as a city in Europe within a region in Europe' (RTPI, 1994: Vol II p. 47).

The example of Birmingham is set out in the RTPI report as an indication of the range of EU activities and memberships that can occur, and of the wide-ranging nature of the work of the European Task Force (RTPI, 1994). This report suggests that the value of an effective European office was shown in the first round of NPCIs and IDOs (see Chapter 5) in the 1980s, and that the Commission has shown itself to be responsive to municipal initiatives.

Consultants

Although the title of this chapter refers to local authorities, one way in which some authorities seek advice on EU affairs is through the appointment of consultants. In addition to the normal range of professional firms, especially in the property development, economic development and environmental assessment fields, there are a number specialising explicitly in EU funding and lobbying advice. Local and regional authorities may appoint such firms to advise on the funding opportunities available to them and perhaps to prepare their funding proposals, or to manage some project, such as an Ouverture project or other form of liaison with the CEE that falls outside the normal range of duties and requires different expertise.

Some firms that have sufficient experience of EU programmes may also succeed in being appointed to manage programmes on behalf of the Commission. The RECITE programme discussed in Chapter 8 provides an example of this.

German local and regional authorities

Although Germany has always actively supported European integration, German municipalities have been less involved in inter-regional co-operation projects and less likely to take the initiative in making proposals to programmes such as RECITE than those of other comparable large member-states. This disparity was so evident that the RECITE office decided to conduct an investigation.

The survey revealed that there were institutional impediments – because the organisation of inter-regional co-operation projects cut across established lines of responsibility and accounting. Another reason cited was lack of multilingual staff, the assumption being that networking was more usually conducted in English or French than German. A more fundamental reason was that German cities in the survey had not developed a clear 'European vision' or perception of how they could contribute, or what benefits it may offer them. A clear preference emerged for co-operation networks within Germany, which may be understandable in the postreunification period since this has itself created a huge demand for advice and transfer of know-how.

It is unlikely that this is the whole explanation, however. The RECITE survey revealed concern that information was not disseminated sufficiently to the municipal authorities. The *Länder* do, of course, receive all EU information and maintain large representations in Brussels, but perhaps because of this the municipalities have tended not to feel the need for establishing their own channels of communication and sources of information on EU opportunities. Certainly one conclusion was that information should be supplied directly to eligible municipalities whenever participation in programmes is invited.

Conclusion

It may seem fanciful to describe it in such terms, but many cities and regional authorities may appear to be conducting their own foreign policy, at least within the EU and often beyond. Networking, lobbying and active participation in EU associations such as Eurocities or RECITE projects should be based on strategic objectives of the authority to which such activities contribute. To the extent that achieving these objectives requires particular alliances and agreements to be reached on joint projects with partners in other countries, this does amount to a form of foreign policy.

The situation may come when every official handles EU aspects of his or her work just as easily as he or she does local and national liaison, regarding the EU as one jurisdiction with several levels of government, rather than a community of different jurisdictions. This would correspond to the ideals of European integration and the rhetoric of a Europe of the regions.

At present, many authorities find EU relations a sufficiently important part of their work to require each of the departments affected to have the relevant expertise available among their staff. This expertise is not yet sufficiently ubiquitous for it to be assumed that every professional officer has this expertise, so it is often necessary to hire specialists or ask some officials to concentrate on adding European liaison to their existing skills as a specialism.

Since local and regional authorities are competing with each other, sometimes very effectively, in seeking advantages for themselves through participation in EU programmes such as the Community Initiatives, there is a danger that the eventual list of approvals lacks the overall spatial coherence that the Commission is seeking and reflects instead the spatial distribution of the effective operators in this competition.

PART IV
Conclusions

15

Future agendas: the new planning scale?

The material reviewed in the previous chapters adds up to a substantial set of spatial policies and programmes, operation of which imposes major responsibilities for those professionally engaged in spatial policy and planning. It is, however, a snapshot at a point in time. The present situation is the product of policy development over many years and is by no means static. Further changes can be expected to continue for a variety of reasons and in response to a variety of pressures.

The growth of spatial policy over the years has taken place sometimes incrementally, sometimes as a result of a more strategic overall review, such as the review leading to the co-ordination of the Structural Funds in 1988. The preparation of the *European Spatial Development Perspective* (*ESDP*; see Chapter 12) is intended to provide the opportunity for strategic development of spatial policy. The opportunity for a much more wide-ranging discussion of the place of spatial strategy within the whole body of EU policy, is presented by the review of the treaties at the 1996 IGC.

The central question is whether it will continue to be necessary in the interests of accuracy to speak of a body of EU spatial policies, or will it become the case that one will be able to speak of EU spatial policy in the singular? In other words, will spatial policy development through the *ESDP* lead to the adoption of a sufficiently robust spatial policy to which other EU policy sectors will cohere? Will this be respected as a context for national, regional and local spatial policy? Secondly, will the process of reviewing the treaties lead to the inclusion of an explicit EU competence over spatial policy to back this up?

The aim of this chapter is to introduce these questions and suggest some issues that spatial planners might need to grasp in the coming years. It also touches on the related questions of an EU urban policy; policy towards town planning; harmonisation and the possible use of Article 130(s); the implications of further enlargement; and of economic and monetary union (EMU). Inevitably, this chapter is therefore speculative and the suggestions contained here liable to be overtaken by events.

The most immediate reason for such uncertainty is of course the IGC due to be convened in 1996. Nevertheless, the possible significance of the IGC for the spatial policy field is discussed in the hope that this will at least offer an indication of issues on which there may be developments to look out for.

Meanwhile, work on spatial planning is continuing, and is outlined first. Particular attention is given below to the implications for spatial policy of the two big topics on the post-IGC political agenda, enlargement and EMU. Finally, the question of a single and coherent EU spatial policy is returned to.

Continuing policy development

Of course, spatial policy development continues prior to the IGC in the directions discussed in Chapter 12 on the basis of existing powers and agreements among the Commission, the Informal Council of Ministers of Regional Planning and the Committee of Spatial Development. During the latter part of 1995 or early 1996, the *ESDP* should reach the final stages of preparation ready for approval by the Informal Council. The Co-operation Network or Observatory of Spatial Research Institutes can embark on its task of monitoring spatial changes throughout the EU and providing research and intelligence advice on which to base the development of policy instruments and implement the spatial policies that may follow from the *ESDP*.

Material from the *Compendium of Spatial Planning Systems and Policies* (see Chapter 12) should become available in 1996. The main purpose of the Compendium is, of course, to make information available to the professionals responsible for planning and development. There is no intention of harmonising planning procedures as such, but it would be naïve to rule out single-market proposals if there is evidence from the research for the Compendium that there are significant non-tariff barriers arising from the different planning systems.

In general, there has been a degree of convergence of style and content of planning systems brought about by practice and the exchange of ideas, not through legislation or harmonisation measures. The Compendium will no doubt accelerate this process. In respect of environmental aspects of planning, convergence may follow from closer scrutiny by the European Environment Agency, but in its first year it has not had much impact on spatial planning.

Meanwhile a series of *Innovative Actions for Regional Development* are to be proposed by DG XVI under Article 10 of the ERDF regulation (Commission, 1995d). These will take the form of calls for proposals under the headings of Inter-regional Co-operation and Regional Economic Innovation, Actions in Spatial Planning and Urban Pilot Actions. The latter two may offer pointers to what the Commission hopes will come out of the IGC.

Actions in Spatial Planning

The spatial planning actions will be intended for contiguous geographical areas, whether within one member-state or transnational, where the integrative or horizontal logic of spatial planning serves to identify the spatial consequences and co-ordinate the implementation of different EU policies on a multisectoral basis. Encouragement will be given to projects which build upon the transnational planning regions such as the Atlantic arc, the Alpine arc and the North Sea basin which were the subject of transnational studies following the *Europe 2000* study of 1991 (see Chapter 6). Another concern in the minds of the Commission is that of environmentally threatened areas such as mountain areas, coastal regions and river basins where spatial planning can address conflicts between environmental protection and tourism and economic development, or address concerns regarding protection against inundation by flooding.

Urban Pilot Actions

A new set of Urban Pilot Projects (see Chapter 11) are to be undertaken during the period 1995–9. These will build upon the experience already obtained, concentrating on disseminating results and employing new ideas gained from them. The emphasis will be on projects tackling problems which are widely shared by many other European cities, and where the public authorities operate in partnership with the private sector. Above all, they must demonstrate the European dimension and not be purely local in their perspective.

The proposals also include a number of measures designed to improve the statistical databases and decision-support systems on which urban policy decisions could be based. Development of new indicators in the form of an 'urban barometer' to measure the quality of life in cities is also envisaged. It is hoped that work along these lines will lead to the development of an urban diagnostics methodology.

Although these Urban Pilot Actions remain within the framework of innovative actions under Article 10, the orientation of the guidelines issued by the Commission would seem capable of being interpreted in terms of laying the foundations for a future EU urban policy, should the IGC decide to propose a legal basis for this.

Review of existing programmes

It should not be forgotten that a number of existing measures are approaching their dates for review. The list of areas eligible for Objective-2 funding under the Structural Funds is to be reviewed so that any revisions can come into effect for the period 1997–9. Discussions with national governments were underway by late 1995 and lobbying and negotiation will no doubt intensify in 1996.

A proposal for revising the 1985 directive on the environmental assessment of projects (see Chapter 10) was agreed in December 1995 by the Environment Council. It then returned to the EP for a second reading under the co-operation procedure, after which formal adoption is expected later in 1996. In addition to seeking to clarify the existing directive in order to make it more consistent in its operation, proposals under consideration aim to extend its scope to include strategic environmental assessment and a wider range of development. Farming, forestry, industrial, energy, transport, tourism, water and waste management and 'town and country planning' (RICS, 1995) projects would require assessment in the proposals under discussion.

In its draft work programme for 1996, the Commission is proposing to launch debates on a number of themes relating to cohesion policy and the regions, including urban issues and the place of cities in spatial policy; regional policy and the information society; and the respective relationships between cohesion and transport and employment policies (EIS, 1995: 26).

The IGC

It was agreed at the Maastricht Summit in 1991 that a further IGC should be convened in 1996 to review the treaties, and specifically to consider any revisions or additions that might be necessary to the Treaty on European Union. The 1996 IGC has not become associated with a particular host city in the way that the 1991 IGC is linked with the name Maastricht. It opened in Turin during the Italian presidency in March 1996, but the timetable is far from being fixed, so it is not yet known which member-state will hold the EU presidency when the time comes for final approval and adoption of the text. The city chosen for this purpose will be the name by which any new treaty is likely to become known. The formal process must start in 1996, but much time will be taken by exchanges between member-states' governments negotiating positions on a very wide range of issues before any proposed new text takes shape. The possibility exists that the process may be deliberately allowed to continue into 1997, delaying agreement on a final text until after the latest date on which the UK general election must take place.

Assuming the IGC succeeds in producing a revised treaty text on which heads of government can agree at a summit sometime in 1997, this would then need to be ratified by all 15 member-states either by referendum or by resolution of parliament. Almost two years elapsed from the Maastricht Summit in December 1991 to the date on which the Treaty on European Union came into force, 1 November 1993. Although it can be anticipated that every effort will be made to avoid the political difficulties faced then in securing popular support for ratification, any new powers over spatial policy or anything else will not necessarily come into effect for some years.

The IGC process had already started in 1995. Many organisations have issued documents setting out their position on issues affecting them, and the lobbying process had already reached a level of some intensity. Meanwhile, the Council has established a 'Reflection Group' of senior politicians from each member-state whose task is to explore the topics which the IGC will be required to address and the range of ideas and initiatives being put forward, identifying points on which general agreement can be expected and those on which conflicting opinions are held.

The published considerations of the Reflection Group (1995) contain little that directly addresses spatial policy, but do indicate some proposed lines of thinking that relate to the overarching issues that will form the context for the future development of EU spatial policy.

Among the issues of institutional reform to be negotiated at the IGC, one is of particular relevance to the theme of spatial policy. The future powers and role of the CoR will undoubtedly figure in the discussions. At one time, there were suggestions that it should not continue to exist, and the committee itself has devoted considerable efforts both to producing well reasoned opinions on issues on which it has expertise to offer and to considering its own future. It is likely that some redefinition of its role and possibly of the numbers and allocation of its seats between member-states will occur.

As far as the spatial planning functions of local and regional government are concerned, much will depend on the extent to which the local and regional government lobbies succeed in pressing their views on the IGC. The UK's Local Government International Bureau, for example, has set out six aims for its campaign under the title '1996 – The Local Government Agenda':

- inclusion in the Treaty of a clear definition of the principle of subsidiarity;
- creation of a legal base for the principle of local self-government;
- revision of the Articles concerning the Committee of the Regions;
- revision of the powers of the EU in the fields of local authority competence to ensure that any necessary European promotion and regulation in these fields is undertaken in partnership with local and regional government;
- inclusion in the Treaty of a legal base for the EU to act in the fields of urban policy and racial equality;
- an explicit requirement for partnership with local and regional government in the operation of the structural funds, particularly if they are to be increased in volume.

(LGIB, 1995)

Although the second and last of these points address specifically UK issues, other points reflect more general concerns of local and regional authorities throughout the EU and are likely to be addressed by the IGC.

Spatial policy

It is perceived by the Commission that there is a need for a clear policy drive and sense of purpose as the IGC and discussions over the nature and extent of any new powers approach. The *Europe 2000* and *Europe 2000+* studies were Commission led, but the European Spatial Development Perspective and the functioning of the Committee of Spatial Development are more intergovernmental in style, giving rise to a sense of loss of central policy focus and thrust. These latter are even seen as deliberately undermining the *Europe 2000+* programme by some observers. In any event, it is likely that the limits have been reached of what is achievable by the intergovernmental mode, given that its consensual mode of operation lacks the decisive lead necessary to reach hard policy conclusions. The Commission takes the view that such a lead is needed in order to achieve policy coherence.

Spatial policy is seen as having a moderating function as a long-term input into the ways other policies are implemented, offering a framework and coherence for them. It offers a means of responding to new political realities and of being a key tool in the redistribution of wealth and adjustment to enlargement. It should be seen as a policy for the whole of the EU and not just selected locations or objective areas. It should embrace environmental as well as economic concerns, for example as a framework for environmental taxation.

At present, the Structural Funds are in danger of taking a form of autopilot operation in which incremental growth of programmes takes place without discernible policy content or overall strategy. Therefore, there is a need to look again at fundamentals in order to defend their interventions against any argument that they are not cost effective, or that they do not support the overall objectives of European integration.

Regional assistance is seen as having a vital role if economic union is to come about, with spatial policy playing an essential framework role and forming a key tool in moving away from nation-states to the concept of a Europe of regions. Therefore, a serious rethink is likely to take place during the period of the IGC of the whole cohesion framework. The aim would be to create a spatial competence in the cohesion or regional titles and an assertion of what the Commission may see as the proper role of the EU (as opposed to member-states) in spatial policy. Above all, the outcome of this process should be greater coherence for EU spatial policy as a whole.

A number of member-states have indicated their desire to add a competence and legal basis for an EU spatial policy to the treaties at the IGC. The Commission also supports this view. This would give the Commission the authority to achieve and maintain coherence of policy, for example by undertaking regional reviews to coincide with reviews of the progress towards attainment of the objectives of the Structural Funds. The Compendium may play a role here. For example, it will provide the basis on which to judge whether any co-ordination of planning timescales and authorisation procedures should be proposed in order to facilitate strategic environmental assessment or implementation of crossborder sections of TENs.

Although completely new powers could be added to the treaties, another possibility would be to amend the existing Article 130(s) to allow it to be used in the way that meets the above purposes. This article does appear to offer the Commission competence in spatial planning if the text is read in French or German, as Chapter 4 pointed out, but the problem is that it is in the wrong part of the treaty, and therefore its use as a legal basis for non-environmental measures may be open to legal challenge. Also, of course, majority voting is seen as a prerequisite if a sufficiently coherent spatial policy is to be adopted.

Urban policy

The LGIB, quoted above, and many other representative bodies of local and regional authorities including Eurocities and the CoR itself are part of a very broadly based campaign to urge the IGC to add an 'urban' title to the treaties, or at least to create an explicit competence and legal basis for an EU urban policy. As Chapter 11 showed, there is a whole range of urban policies, and a new set of Urban Pilot Projects is to be launched in 1996. Whether this will continue to rely upon Article 10 and Community Initiative funding under ERDF, or whether it will become possible to talk of urban policy as an EU policy sector in its own right, with an independent existence based on treaty powers, remains to be seen.

Enlargement

The next round of enlargement negotiations are due to begin directly after the IGC. In fact, the pressure of expectation from the countries concerned places a limit on the extent to which the IGC timescale can be drawn out. The September 1995 progress report of the Reflection Group recognises that the next enlargement 'represents both a moral imperative and a major opportunity for Europe' (Reflection Group, 1995: 3), but also that it poses major challenges to the institutional structure of the EU and to several policy sectors because of the large number of countries involved and their political, economic and environmental situation.

There are three groups of countries with whom accession negotiations may take place. First in line are the four Visegrad countries (see Chapter 2) who have agreed to act in concert in their accession negotiations: the Czech Republic, Hungary, Poland and Slovakia. Secondly, there are the other central European countries that have, or expect to have, concluded 'Europe agreements' with the EU: Bulgaria, Estonia, Latvia, Lithuania, Romania and Slovenia. Thirdly are two small Mediterranean countries, Cyprus and Malta. This leaves Turkey, whose candidacy long predates all the other potential new members. Also waiting in the wings are Albania, Croatia and other parts of the former Yugoslavia.

The IGC itself will not consider applications for membership, but it will have to consider institutional changes, for example to the structure of the Commission; the voting system in the Council of Ministers; and the rotating presidency – which will be necessary if an EU of 25-plus member-states is to become a reality.

The present basis for several policy sectors will also need to be re-examined. Spatial policy will be central to this process, since the whole spatial structure of the EU will once again be profoundly changed. Extension of present policy thinking to the CEE has possibly progressed most in respect of the programme of TENs. A conference was convened in Prague in October 1995 by the Council of Europe and the Commission to consider the question of spatial planning for the enlarged EU. Using as its basis the *Europe 2000+* study, it examined three main themes:

1. Spatial integration of the CEE.
2. Support for their transformation through the development of infrastructure, urban networks, industrial conversion and sustainable use of resources.
3. Exchange of information on regional planning and development policies between regional and national authorities.

Spatial designation of priorities for targeting financial support for economic development and environmental improvement will be a key issue in enlargement negotiations. However, if the level of regional aid offered to new member-states from central Europe were to match the levels currently on offer to the poorer regions of the EU of 15 member-states, this would require a massive addition to the EU budget which is unlikely to find ready approval. Instead, it is anticipated that a long transition period would be proposed for the accession countries, not least in order to keep the pressure to fund an enlarged budget down to manageable proportions.

Whereas the cut-off point for participation in the Cohesion Fund is a national per capita GDP of 90% of the EU average, and for designation as an Objective-1 region is a per capita GDP of 75% of the EU average at the NUTS-2 regional level, the average GDP per capita of the Europe agreement countries is around 36% of the EU average. Clearly, the present basis by which eligibility for the Cohesion Fund and Objective 1 is determined will eventually have to be revised. It will not simply be a question of adding a new and higher priority for former communist countries. The present basis for EU funding programmes requires cofinancing from national sources, but any similar requirement for an enhanced objective status for new central European member-states would run into the problem that they simply would not have the resources to provide matching funds for necessary projects.

Environmental degradation and pollution in the former communist countries has in some regions been vastly more severe than anything experienced in western Europe. In addition to investment in infrastructure and economic development projects, massive investment in environmental improvement will be necessary over a long period before anything approaching the current standards set by EU environment policy can be attained.

Although there is considerable political support for enlargement throughout the EU, the implications of eastward enlargement are not lost on existing member-states, especially those that are currently the chief beneficiaries of the Cohesion and Structural Funds. There is a general sense of moral imperative to consolidate the economic and political reforms of the former communist countries and to minimise the risk of future instability and threats to security by locking them into the west European or EU economy. Some existing member-states (including those currently on the eastern periphery, such as Finland and Austria) believe that they stand to benefit from the consequent changes in the spatial structure of the EU. Their location and postwar role as bridge-states in a divided Europe puts them in a good position to play a key role in an enlarged EU. Other peripheral member-states, especially the poorer cohesion countries on the southern and western periphery, may feel they have more to lose and less to gain.

Gower argues that there is a real danger of 'rival peripheries' emerging from the different parts of the EU which could cause tension in the future (Gower, 1995). A well argued challenge to the conventional wisdom concerning the economic disadvantages faced by peripheral regions is presented by Chisholm (1995). Nevertheless, it is recognised in several parts of the existing periphery (such as the Objective-6 areas of Finland and the least poor parts of the Cohesion Fund countries) that they need to make the most of their present status while they retain the highest priority, because they can no longer expect the highest priority under any post-enlargement spatial policy. The Reflection Group recognises the need to offer some protection to the weakest parts of the present EU because it is determined that enlargement should proceed successfully, and refers to the suggestion that a provision could be incorporated into the revised treaty dealing with support for the 'outermost regions of Member States of the Union' (Reflection Group, 1995: 43).

EMU

In a general sense, the link between spatial policy and economic and social cohesion is clear from the terms of reference of many spatial programmes, for example in the priorities of Community Support Frameworks and objectives of Community Initiatives funded through the Structural Funds. This link also underlies *Europe 2000+* and the *ESDP*.

However, there is another possible imperative for the development of EU spatial policy. A more explicit link between the objective of achieving a coherent spatial policy and that of EMU or the single currency has not been articulated but would seem to be a possible explanation for the sense of purpose with which it is pursued and broad political support in certain member-states. The link is based on the argument that effective spatial policy is needed if EMU is to come about, as a basis on which to provide counterbalancing support for those regions and member-states that are adversely affected by EMU for reasons of loss of competitiveness. Compensation through exchange rate fluctuation and national fiscal policy response would not of course be available after EMU. One way in which any compensation could be provided is through regional assistance programmes such as the Cohesion Fund and those directed at the spatially targeted objectives (Objectives 1, 2, 5(b) and 6) of the Structural Funds (see Chapter 7).

A close observer of political trends in the EU that affect local and regional interests notes that, in 1992, 'the concept [of economic and social cohesion] acquired a new dimension as compensation for the economic impact of moves towards Economic and Monetary Union'. Hence the creation of the Cohesion Fund at Maastricht. He goes on to argue that

> this concept will be revisited again at the 1996 Intergovernmental Conference (IGC), which will take place against the background of moves towards a single currency. There are signs that the Commission will seek to link the concept to 'aménagement du territoire' – the strategic planning of the use of the European space.
>
> (Gallacher, 1995b: 7)

It is normally assumed to be absurd that the inclusion of 'town and country planning' or *aménagement du territoire* in Article 130(s) of the Maastricht Treaty is subject to national veto and not QMV since matters of 'town and country planning' and 'land use' are not normally considered to be matters of vital national interest – but perhaps this is not so far wide of the mark. Likewise, one could speculate about the UK Prime Minister's seemingly inadvertent linkage of planning policy with the big political issues, quoted in Chapter 1.

Concluding remarks

One question posed by those who see no role for EU spatial policy is why, if the EU is modelled on the USA as the United States of Europe, is there no spatial policy for the USA or USA 2000? This begs the question of whether there ought to be and would have been if the USA had a European spatial structure, densities and mentalities. Although there is little evidence of continental spatial policy development in North America, it is possible that people will come to recognise and advocate that NAFTA should develop a spatial competence. It is also possible that some people will point to the spatial legacy of changing industrial technologies, abandonment of large areas of the older industrial 'rustbelt' cities in the USA, and to tensions created by NAFTA, in order to argue that EMU could have similar effects unless compensatory policy instruments of a spatially targeted type are adopted.

It is possible that it will all go away. *Europe 2000+* has no legal status and could perhaps suffer the fate of the *Green Book on the Urban Environment.* But with the CoR, the EP and the new Commissioner all calling for more integrated spatial planning, seeing it as a key tool in the move away from nation-states to the Europe of the regions, this fate seems less likely. The objective of arriving at a single, coherent, EU spatial policy giving strategic direction to all the elements and sectors discussed in earlier chapters, in support of the overall goals of the EU as a whole, is firmly adhered to by those responsible within the Commission.

It is taken very seriously by the CoR because the development of spatial policy is seen by the CoR as being interlinked with its own future existence, standing and *raison d'être* following the IGC. If spatial policy develops in the way its protagonists hope, this will then help to ensure that the CoR has an essential role (BfLR, 1995; Williams, 1996). Preparation of its opinion on the *Europe 2000+* study was seen by the CoR as one of its most important tasks in 1995 – important for the committee's own reputation as much as for their substantive comments. This did not go unnoticed by the Commissioner, Monika Wulf-Mathies: 'European land-use policy is particularly important to me . . . I therefore welcome the fact that, in your Opinion on Europe 2000+, you drew attention to the importance of European land-use policy' (CoR, 1995b: 3).

In its final debate in its opinion in July 1995, the contents of *Europe 2000+* were hardly examined, the main debate being about how far the CoR could lay claim to manage and direct EU spatial policy, and on proposals to argue to the IGC that the CoR should elaborate spatial policy. The conclusion was that it will not go away, and also that it does in fact tie up with some of the 'big' politics concerning powers of EU institutions, EMU, sovereignty, the rhetoric of a Europe of the regions and eastward enlargement.

So it seems reasonable to draw the positive conclusion that spatial policy will continue to be developed and play a role in giving coherence to, and coordinating the actions of, EU sectoral policies. Further support for this view could be taken from the fact that the EU presidencies for 1997, The Netherlands and Luxembourg, are both countries which have in the past devoted considerable resources to the development of European spatial policy.

Above all, spatial policy formulation at the European scale is a challenge to the imagination. Too many policy-makers show themselves to be set in their ways and inhibited, or nationally bounded, in their thinking. European integration requires not only new governmental structures and physical infrastructure links but also new mental maps and removal of Cartesian inhibitions. It is necessary for policy-makers to learn to think European, and for educators to develop the capacity for this mode of thinking in their students. It is hoped that readers will be equipped to develop new ways of thinking not only to accommodate the EU spatial scale but also to make the mental leap required to conceptualise a continental jurisdiction and planning subject.

The next generation of spatial planners may be brought up on the story of EMU, the Blue Banana and the Bunch of Grapes!

Appendix I: Guide to information sources

The most important resource is the network of European Documentation Centres, generally located within university libraries in all the main cities and regional centres throughout the EU.

All designated European Documentation Centres receive copies of all EU publications in the language of the country in which it is located. This means

- all Commission publications, Council of Ministers documents, studies, bulletins and reports;
- the *Official Journal*;
- opinions of the EP, CoR and Ecosoc;
- publications of the other EU institutions;
- newsletters and publicity material from the EU institutions; and
- statistical information published by Eurostat, the EU's statistical office. For guidance on the latter, see Ramsay (1992).

A lot of information is now available on CD-ROM and on the Internet.

All European Documentation Centres are open to the general user, and are not confined to members of the university in which they are situated.

The Commission makes a great deal of material available free. Some documents can be obtained on written request to the Commission or to the Commission's offices in each national capital city. Near the Commission's offices in Brussels, there is an Information Shop from which copies of documents available free can be obtained.

The volume of official publication is so enormous (the *OJ* is published daily, for example) that organisations and individuals concerned only with certain specific EU topics often need guidance on what to look out for, when something on that topic is issued and where to find it. Fortunately, there are several publications, advisory services and firms of consultants which offer such guidance.

For up-to-the minute news, the newsletters of the Brussels-based press agencies, such as *Europe* and *European Report*, which are published most days, are most useful. Several organisations prepare their own EU circulars and newsletters, sometimes for internal use only, sometimes for wider circulation or publication. These are often published on a subscription basis. A particularly useful one for local authorities in the UK is the *European Information Service*, a monthly publication of the UK Local Government International Bureau in London. A more comprehensive current-awareness bulletin is *European Access*, published bimonthly by Chadwyck-Healey of Cambridge.

For more general guidance on the institutions, who does what and how to make contact, several directories are on offer. The American Chamber of Commerce *EU Information Handbook* is very well presented and easy to use (AmCham, 1995). *Vacher's European Companion and Consultants Register* (Vachers Publications, Berkhamstead) is published quarterly and is also very useful. Another source is *The Europe Directory: A Research and Information Guide* (3rd edn 1995, HMSO, London). The Commission also publishes its own *Interinstitutional Directory*.

At a more introductory level, Budd and Jones (1991), Owen and Dynes (1993) and Noel (1994) and textbooks such as Nugent (1991) are recommended. Acronyms and abbreviations themselves can sometimes be a problem, for which a valuable glossary is Ramsay (1994).

For guidance on more scholarly work on the EU, Evans (1995) is a helpful bibliography, and Lodge (1993) covers a wide range of material. Several planning journals regularly carry articles on EU topics, and there are now some with explicit interests in EU spatial policy including *European Planning Studies* and *European Spatial Research and Policy*, while the *Journal of Common Market Studies* concentrates more on political analysis.

Appendix II: Guide to educational programmes

This appendix aims to provide a simple introduction to the programmes that may help readers, particularly students, who want to study or have work experience in another member-state.

Many students have been able to take advantage of the ERASMUS programme. ERASMUS is the acronym for European Community Action Scheme for the Mobility of University Students. It came into being in 1987 with the ambition that 10% of all students should have the opportunity of some study within another member-state of the EU. Although these numbers have not been achieved, it has proved very popular and successful, becoming a standard feature of university life.

For the academic year 1997–8 it is to be replaced by SOCRATES (not an acronym), which will offer an extended range of opportunities. An essential component will be the institutional contracts between individual universities and the SOCRATES office of DG XXII in Brussels. Anyone wishing to incorporate an exchange within their own university studies should ask whether this will be available at their chosen university.

ERASMUS and SOCRATES are intended to be reciprocal programmes for the whole EU. However, for language reasons, the UK and Ireland attract more than they send abroad, since English is by far the first choice of second language among students from the rest of the EU.

SOCRATES includes all the EEA countries (i.e. the EU plus EFTA, but not Switzerland). It is expected that it will extend to the applicant countries of central Europe before long, and some years before any date of accession that may be agreed.

Meanwhile, the TEMPUS programme provides for student exchanges and teaching assistance for the former communist countries. Unlike ERASMUS, this is not intended to be a programme of reciprocal exchanges. There are opportunities under TEMPUS for EU students to study in CEE universities, but the numbers coming westwards is much greater. Research students and research staff of universities may be able to participate in the Human Capital and Mobility Programme, which is designed to strengthen Europe's scientific human resource.

Work experience in another EU country is another form of mobility that is becoming popular. The COMETT programme (Community Programme in Education and Training for Technology) has funded this through 'university–enterprise training partnerships' (UETP) which organise placements or *stages* (the French term is coming into general use) in countries other than the student's home country. In 1995–6 COMETT is being replaced by Leonardo (not an acronym) which will extend the range of possibilities.

Brussels is a city of *stagières*, i.e. students undertaking a break in their studies to gain practical experience. Many employers, including many lobby organisations and representations of local and regional authorities, will offer some months' work experience to *stagières*. Sometimes this will be paid work; however, sometimes unpaid work experience is all that may be available. Anyone who has the opportunity of a grant through Leonardo or other sponsorship which provides at least minimum living costs may be able to find such employment. Prospective employers often look out for *stagières*, some being really quite dependent on them, and varied and interesting work may be offered.

The élite of the *stagières* are those accepted for the Commission's own scheme. The competition for places on the *stagière* scheme is intense. Enquiries should be made to the Bureau des Stages at the Commission or the recruitment officer of national offices of the Commission. University careers services can often offer advice or obtain application forms.

For all purposes, written enquiries to the Commission should be addressed to the appropriate DG or section at the postal address of the Commission, rue de la Loi 200, B-1049 Brussels, Belgium.

References and further reading

Acosta, R. and Renard, V. (1993) *Urban Land and Property Markets in France*, UCL Press, London.

ADC (1985) *EEC Policies: Views on Issues of Concern to Districts*, Association of District Councils, London.

Albrechts, L. (1995) Emerging European Planning. A First Critical Reflection. Paper presented to 9th AESOP Congress, University of Strathclyde, Glasgow, 19 pp.

Albrechts, L., Kunzmann, K.R., Motte, A. and Williams, R.H. (1990) *Towards a European Core Curriculum in Planning Education*, AESOP, Fakultät Raumplanung, University of Dortmund, Dortmund.

Alden, J. and Boland, P. (eds) (1996) *Regional Development Strategies: A European Perspective*, Jessica Kingsley, London.

AmCham (1995) *EU Information Handbook 1995/1996*, EU Committee of the American Chamber of Commerce in Belgium, Brussels.

Anselin, M. (1984) Belgium, in Williams, R.H. (ed.) *Planning in Europe*, Allen & Unwin, London, pp. 63–72.

Ave, G. (1996) *Urban Land and Property Markets in Italy*, UCL Press, London.

Bangemann, M. (1994) *Europe and the Global Information Society*, DG XIII, Brussels.

BfLR (1995) 'The EU Committee of the Regions' theme issue of *Informationen zur Raumentwicklung*, November.

Blacksell, M. (1981) *Post-War Europe: A Political Geography*, Hutchinson, London.

BMBau (1992) *Spatial Planning Concept for the Development of the New* Länder, Bundesbauministerium, Bonn.

BMBau (1993) *Raumordnungspolitischer Orientierungsrahmen Leitbilder für die räumliche Entwicklung der Bundesrepublik Deutschland*, also published as *Guidelines for Regional Planning: General Principles for Spatial Development Planning in the Federal Republic of Germany*, Bundesbauministerium, Bonn.

BMBau (1993b) *Network of Spatial Research Institutes in Central and Eastern Europe*, Newsletter No. 2, Bundesbauministerium, Bonn.

BMBau (1994a) *Spatial Planning Policies in a European Context*, publicity leaflet, Bonn.

BMBau (1994b) *European Spatial Planning: Results of the Meeting, Informal Council of Spatial Planning Ministers, Leipzig*, September, Bundesbauministerium, Bonn.

BMBau (1995) *Principles for a European Spatial Development Policy*, Bundesbauministerium, Bonn.

Boland, P. (1995) Objective 1 status: economic and urban regeneration in Merseyside. Paper presented to 9th AESOP Congress, University of Strathclyde, Glasgow.

Bongers, P. (1990) *Local Government and 1992*, Longman, Harlow.

Borchardt, K.-D. (1995) *European Integration*, European Documentation OOPEC, Luxembourg.

Brundtland, G.H. (1987) *Our Common Future: World Commission on Environment and Development* (the Brundtland Commission), Oxford University Press, Oxford.
Brunet, R. (1989) *Les villes européennes*, La Documentation française, RECLUS, DATAR, Paris.
Budd, S.A. and Jones, A. (1991) *The European Community: A Guide to the Maze* (4th edn), Kogan Page, London.
Cecchini, P. (1988) *The European Challenge 1992: The benefits of a Single Market*, Wildwood House, Aldershot.
Cheshire, P. and Hay, D. (1989) *Urban Problems in Western Europe*, Unwin Hyman, London.
Chisholm, M. (1995) *Britain on the Edge of Europe*, Routledge, London.
CLES (1992) *Building the New Europe: Baltic Gateways Project, Phase I: Interim Report*, May, CLES, Manchester.
CoE (1984) *European Regional/Spatial Planning Charter*, Council of Europe, Strasbourg.
CoE (1993) *Convention on Civil Liability for Damage Resulting from Activities Dangerous to the Environment. European Treaty Series* 150, Lugano.
Cole, J. and Cole, F. (1993) *The Geography of the European Community*, Routledge, London.
Commission (1973) *Report on the Regional Problems of the Enlarged Community* (COM (73) 550), Brussels.
Commission (1979) *Memorandum on the Role of the Community in the Development of Transport Infrastructure* (COM (79) 550), Brussels.
Commission (1980a) *First Periodic Report on the Social and Economic Condition in the Regions of the Community* (COM (80) 816), Brussels.
Commission (1980b) Proposal for a Council directive concerning the assessment of the environmental effects of certain public and private projects, *Official Journal* C, 169, Brussels.
Commission (1982) Proposal to amend the proposal for a Council directive concerning the assessment of the environmental effects of certain public and private projects, *Official Journal* C, 110, Brussels.
Commission (1985) *Completing the Internal Market*. White Paper, June.
Commission (1987) *The State of the Environment in the European Community 1986*, DG XI, Brussels.
Commission (1989) *Proposal for a Council Directive on Civil Liability for Damage caused by Waste* (COM (89) 282), Brussels.
Commission (1990a) The European Community and German reunification, *Bulletin of the European Communities*, Supplement 4/90, Luxembourg.
Commission (1990b) *Green Book on the Urban Environment* (COM (90) 218), Brussels.
Commission (1991) *Europe 2000: Outlook for the Development of the Community's Territory*, DG XVI, Brussels.
Commission (1992a) *From the Single Act to Maastricht and Beyond: The Means to Match our Ambitions* (COM (92) 2000), Brussels.
Commission (1992b) *Fifth Action Programme on the Environment 1992–2000 'Towards Sustainability'* (COM (92) 23), Brussels.
Commission (1992c) *Green Paper on the Impact of Transport on the Environment. A Community Strategy for 'Sustainable Mobility'* (COM (92) 46), Brussels.
Commission (1992d) *Commission Communication on the Future Development of the Common Transport Policy* (COM (92) 494), Brussels.
Commission (1992e) *Proposal for a Council Decision on the Creation of a Trans-European Road Network and on the Creation of a European Inland Waterway Network* (COM (92) 321 final), Brussels.
Commission (1993a) *Green Paper on Remedying Environmental Damage* (COM (93) 47), Brussels.

Commission (1993b) *Growth, Competitiveness and Employment* (COM (93) 700), Brussels.
Commission (1993c) *Communication from the Commission to the Council, the European Parliament and the Economic and Social Committee on Energy and Economic and Social Cohesion* (COM (93) 645), Brussels.
Commission (1993d) *Commission Communication on the Development of Guidelines for the Trans-European Network System* (COM (93) 701), Brussels.
Commission (1994a) *Europe 2000+: Co-operation for European Territorial Development*, DG XVI, Brussels.
Commission (1994b) *Competitiveness and Cohesion Trends in the Regions. Fifth Periodic Report on the Social and Economic Situation and Development of the Regions in the Community*, DG XVI, Brussels.
Commission (1994c) *List of Areas Eligible for Object 2*. Fact sheet Annex, DG XVI, Brussels.
Commission (1994d) *Guide to the Community Initiatives 1994–99*, Brussels.
Commission (1994e) *Proposal for a Council Regulation Laying Down General Rules for the Granting of Community Financial Aid in the Field of Trans-European Networks* (COM (94) 62), Brussels.
Commission (1994f) *Interim Review of Implementation of the European Community Programme of Policy and Action in Relation to the Environment and Sustainable Development 'Towards Sustainability'* (COM (94) 453), Brussels.
Commission (1995a) *Preliminary Draft General Budget of the European Communities for the Financial Year 1996* (SEC (95) 850-en), Brussels.
Commission (1995b) *Single Programming Document 1995–9, Objective 6, Finland*, DG XVI, Brussels.
Commission (1995c) *The Trans-European Transport Network*, DG VII, Brussels.
Commission (1995d) *Innovative Actions for Regional Development*, DG XVI, Brussels.
CoR (1995a) *Opinion of Commission 5 on the Commission Communication on Co-operation for European Territorial Development* (CdR 52/95, 12 June), Committee of the Regions, Brussels.
CoR (1995b) *Minutes of the 7th Plenary Session of the Committee of the Regions* (Appendix II, CdR 149/95), Committee of the Regions, Brussels.
DATAR (1990) *Vingt Technopoles: un premier bilan*. La documentation française, Paris.
Davies, H.W.E. (1989) *Planning Control in Western Europe*, HMSO, London.
Dawson, J. (1992) European city networks: experiments in transnational urban collaboration, *The Planner*, Vol. 78, no. 1, pp. 7–9.
de Brabander, G. and Verhetsel, A. (1995) Brussels, in Berry, J. and McGreal, S. (eds) *European Cities, Planning Systems and Property Markets*, E. & F.N. Spon, London, pp. 65–81.
Dieterich, H., Dransfeld, E. and Voss, W. (1993) *Urban Land and Property Markets in Germany*, UCL Press, London.
Dieterich, H., Williams, R.H. and Wood, B. (1993/6) *European Urban Land and Property Markets Series* (6 vols), UCL Press, London.
DoE (1994) *Paying for Our Past*, Department of the Environment and Welsh Office consultation paper, London.
Dransfeld, E. and Voss, W. (1993) *Funktionswiese städtischer Bodenmärkte in Mitgliedstaaten der Europäischen Gemeinschaft – ein Systemvergleich*, Bundesbauministerium, Bonn.
du Granrut, C. (1995) *Europe, le temps des Régions*, Librairie Generale du Droit et la Jurisprudence, Paris.
Dutt, A.K. and Costa, F. (1992) *Perspectives on Planning and Urban Development in Belgium*, Kluwer, Dordrecht.
EC Council (1992) *Treaty on European Union*, Office for Official Publications of the European Communities, Luxembourg.
ECTP (1993) *What is the European Council of Town Planners?* ECTP, Brussels.

EEB (1994) *EEB Twentieth Anniversary*, European Environment Bureau, Brussels.

EIS (1994a) *Charles Gray – a Man for all Regions, European Information Service* 146, Local Government International Bureau, London, pp. 3–7.

EIS (1994b) *What's on the Environmental Agenda? European Information Service* 155, Local Government International Bureau, London.

EIS (1994c) *City Studies. European Information Service* 151, Local Government International Bureau, London.

EIS (1994d) *Make Towns not War! – European Conference on Sustainable Cities & Towns, European Information Service* 150, Local Government International Bureau, London, pp. 25–9.

EIS (1995) *1996 Commission Work Programme – a Preview. European Information Service* 164, Local Government International Bureau, London.

EP (1980) *Report on the Memorandum of the Commission on the Role of the Community in the Development of Transport Infrastructure*. Document 1–601/80, 15 December, Luxembourg.

EP (1983) *Report Drawn up on Behalf of the Committee on Regional Policy and Regional Planning on a European Regional Planning Scheme* (EP doc. 1–1026, 21 November), Luxembourg.

EP (1995) *Report on the Commission Document on* Europe 2000+, *Committee on Regional Policy* (A4–0147/95, PE 211.942, 15 June), Brussels.

Evans, E. (1995) *A Selective Bibliography of Books on European Integration 1990–1994. UACES Information Guide* 25, UACES, London.

Fairclough, A.J. (1983) The Community's environmental policy, in Macrory, R. (ed.) *Britain, Europe and the Environment*, ICCET, Imperial College, London, pp. 19–34.

Faludi, A. and van der Valk, A. (1994) *Rule and Order: Dutch Planning Doctrine in the Twentieth Century*, Kluwer, Dordrecht.

Fayman, S., Metge, P., Spiekermann, K., Wegener, M., Flowerdew, T. and Williams, I. (1995) *The Regional Impact of the Channel Tunnel: Qualitative and Quantitative Analysis, European Planning Studies*, Vol. 3 no. 3, pp. 333–56.

Fudge, C. (1993) *Fragmentation and Progress – Urban Environment Expert Group Focusses on Sustainable Cities. European Information Service* 143, Local Government International Bureau, London.

Gallacher, J. (1995a) *The Committee of the Regions: An Opportunity for Influence. Special Report* 3, Local Government International Bureau, London.

Gallacher, J. (1995b) *Economic Cohesion, the Structural Funds and the Enlarged Europe of the 21st Century. European Information Service* 161, Local Government International Bureau, London.

George, S. (1991) *Politics and Policy in the European Community* (2nd edn), Oxford University Press, Oxford.

Glasson, J. (1995) *Regional Planning and the Environment: Time for a SEA Change, Urban Studies*, Vol. 32, no. 4/5, pp. 713–31.

Glasson, J., Therivel, R. and Chadwick, A. (1994) *Introduction to Environmental Impact Assessment*, UCL Press, London.

Government of Ireland (1993) *Ireland National Development Plan 1994–1999*, Stationary Office, Dublin.

Gower, J. (1995) *Crisis or Opportunity? EU Enlargement to Central and Eastern Europe. Discussion Paper of the Jean Monnet Group of Experts*, Centre for European Union Studies, University of Hull, Hull.

Gray, C. (1994) *Charles Gray – a Man for All Regions. European Information Service* 146, Local Government International Bureau, London.

Group of Focal Points (1994) *Vision and Strategies around the Baltic Sea 2010: Towards a Framework for Spatial Development in the Baltic Sea Region*, The Baltic Institute, Karlskrona.

Haigh, N. (1989) *EEC Environmental Policy and Britain*, Longman, Harlow.

Hardy, S., Hart, M., Albrechts, L. and Katos, A. (eds) (1995) *An Enlarged Europe: Regions in Competition?* Jessica Kingsley, London.

Healey, P. and Williams, R.H. (1993) European urban planning systems: diversity and convergence (Urban Studies annual review 1993), *Urban Studies*, Vol. 30, no. 4/5, pp. 699–718.

Hennekam, B.M.J. (1994) *Second Outline Structural Plan for the Benelux. Informational Brochure on the Agreement*, Benelux Economic Union, Brussels.

Holland, M. (1994) *European Integration: From Community to Union*, Pinter, London.

Holliday, I., Marcou, G. and Vickerman, R. (1991) *The Channel Tunnel. Public Policy, Regional Development and European Integration*, Belhaven, London and New York.

Hughes, I. (1994) *Towards a European Urban Policy? European Information Service* 155, Local Government International Bureau, London.

John, P. (1994) *Do You Need a Brussels Office? European Information Service* 151, Local Government International Bureau, London, pp. 7–9.

Kalbro, T. and Mattsson, H. (1995) *Urban Land and Property Markets in Sweden*, UCL Press, London.

Kramer, H. (1993) The EC's response to the 'New' eastern Europe, *Journal of Common Market Studies*, Vol. 31, no. 2, pp. 213–44.

Krautzberger, M. and Selke, W. (1994) *Auf dem Wege zu einem Europäischen Raumentwicklungskonzept. Die Öffentliche Verwaltung*, Heft 16, pp. 685–90.

Kruijtbosch, E.D.J. (1987) Physical planning in the Benelux – a successful experiment, in Angenent, J.J.M. and Bongenaar, A. (eds) *Planning Without a Passport: The Future of European Spatial Planning. Netherlands Geographical Studies* 44, Royal Dutch Geographical Society KNAG, Amsterdam, pp. 144–54.

Kunzmann, K.R. (1990) *Durch Wiese und Wald in die europäische Stadtpolitik – zum Grünbuch der Kommission der Europäischen Gemeinschaft über die städtische Umwelt. IRPUD Arbeitspapier* 81, Fakultät Raumplanung, University of Dortmund, Dortmund.

Kunzmann, K.R. (1992) Zur Entwicklung der Stadtsysteme in Europa. *Mitteilungen der Österreichischen Geographischen Gesellschaft* 134, pp 25–50.

Kunzmann, K.R. and Wegener, M. (1991) *The Pattern of Urbanisation in Western Europe 1960–1990. Berichte aus dem Institut für Raumplanung* 28, Universität Dortmund, Dortmund.

LGIB (1995) *International News*, Local Government International Bureau, London, July.

Liaison Committee (1985) *International Agreement and Declaration by the National Institutes and Associations of Professional Town Planners within the European Economic Community*, RTPI, London, 8 November.

Lodge, J. (ed.) (1993) *The European Community and the Challenge of the Future*, Pinter, London.

Lodge, J. (ed.) (1995) *Discussion Papers of the Jean Monnet Group of Experts*, Centre for European Union Studies, University of Hull, Hull.

Logie, G. (1986) *Glossary of Planning and Development. International Planning Glossary 5*, Elsevier, Amsterdam.

Macrory, R. (ed.) (1983) *Britain, Europe and the Environment*, ICCET, Imperial College, London.

Macrory, R. and Hollins, S. (1995) *A Sourcebook of European Community Environmental Law*, Clarendon Press, Oxford.

Masser, I., Sviden, O. and Wegener, M. (1992) *The Geography of Europe's Futures*, Belhaven, London.

Mazey, S. and Richardson, J. (eds) (1993) *Lobbying in the European Community*, Oxford University Press, Oxford.

McDonald, R. (1995) TENs cost hundreds of billions, *EIU European Trends* (1st quarter), Economist Intelligence Unit, London, pp. 1–12.

Merritt, G. (1991) *Eastern Europe and the USSR*, OOPEC, Luxembourg.

Meyer, P.B., Williams, R.H. and Yount, K.R. (1995) *Contaminated Land Reclamation, Redevelopment and Reuse in the United States and the European Union*, Edward Elgar, Cheltenham.

MoE (1992) *Denmark Towards the Year 2018: The Spatial Structuring of Denmark in the Future of Europe*, Ministry of the Environment, Copenhagen.

MoE (1993) *The Oresund Region – a Europole*, Ministry of the Environment, Copenhagen.

MoE (1994) *Spatial Planning in Denmark*, Ministry of the Environment, Copenhagen.

Monnet, J. (1978) *Memoirs*, Collins, London.

Nadin, V., Shaw, D. and Westlake, T. (1995) Convergence or divergence? A comparative view of European planning systems. Paper presented to the 9th AESOP Congress, University of Strathclyde, Glasgow.

NEA (1994) North of England Assembly information leaflet, Newcastle upon Tyne.

Needham, B., Koenders, P. and Kruijt, B. (1993) *Urban Land and Property Markets in The Netherlands*, UCL Press, London.

Nevin, E. (1990) *The Economics of Europe*, Macmillan, Basingstoke.

Nijkamp, P. (1993) Towards a Network of Regions: The United States of Europe, *European Planning Studies*, Vol. 1, no. 2, pp. 149–68.

Nijkamp, P. and Perrels, A. (1994) *Sustainable Cities in Europe*, Earthscan, London.

Noël, E. (1994) *Working Together – the Institutions of the European Community*, European Documentation, Luxembourg.

Nugent, N. (1991) *The Government and Politics of the European Community* (2nd edn), Macmillan, Basingstoke.

Owen, R. and Dynes, M. (1993) *The Times Guide to the Single European Market*, Times Books, London.

Parkinson, M. (1992) City links, *Town and Country Planning*, Vol. 61, no. 9, pp. 235–6.

Peizerat, C. (1995) Discourse analysis for European planning and property development research: the example of business parks in England and France. Paper presented to the 9th AESOP Congress, University of Strathclyde, Glasgow.

Pinder, D. (1983) *Regional Economic Development and Policy Theory and Practice in the European Community*, Allen & Unwin, London.

Pinder, J. (1991) *European Community: The Building of a Union*, Oxford University Press, Oxford.

Ramsay, A. (1992) *Eurostat Index* (5th edn), Capital Planning Information, Stamford.

Ramsay, A. (1994) *Eurojargon – a Dictionary of European Union Acronyms, Abbreviations and Sobriquets* (4th edn), Capital Planning Information, Stamford.

RECITE (1995a) *Urban Pilot Projects: Second Interim Report on the Progress of Urban Pilot Projects Funded by the European Regional Development Fund*, RECITE Office, DG XVI, ECOTEC, Brussels.

RECITE (1995b) *Interregional Cooperation Projects: A Guide for Project Management*, RECITE Office, DG XVI, ECOTEC, Brussels.

Reflection Group (1995) Progress report from the Chairman of the Reflection Group on the 1996 Intergovernmental Conference, Madrid.

RETI (1992) *Study of Derelict Land in the ECSC. Vol. A: Summary and Recommendations*, RETI, Brussels.

RICS (1995) *European Alert*, RICS/EACS, Brussels, October.

Roberts, P., Hart, T. and Thomas, K. (1993) *Europe: A Handbook for Local Authorities*, CLES, Manchester.

Robson, B. (1992) Networking of cities, *Town and Country Planning*, Vol. 61, no. 9, pp. 234–5.

Rodriguez-Bachiller, A. (1988) *Town Planning Education*, Avebury, Aldershot.

Rosciszewski, M. (1993) *Poland and the New Political and Economic Order in Europe*, Institute of Geography, Polish Academy of Sciences, Warsaw.

RPD (1991) *Perspectives in Europe*, Rijksplanologische Dienst, Den Haag.

RTPI (1991) *Guidance Note on Initial Professional Education Programmes in Planning*, Royal Town Planning Institute, London.

RTPI (1994) *The Impact of the European Community on Land Use Planning in the United Kingdom*, Royal Town Planning Institute, London.

Schuster, G. (1950) *Report of the Committee on Qualifications of Planners (Chairman Sir George Schuster)* (Cmd 8059), HMSO, London.

Scott, Andrew (1993a) Financing the Community: The Delors II package, in Lodge, J. (ed.) *The European Community and the Challenge of the Future*, Pinter, London, pp. 69–88.

Scott, Adam (1993b) Baltic gateways: building the new Europe, *Business Review North*, Vol. 5, no. 2, pp. 24–9.

Shackleton, M. (1993) The budget of the EC: structure and process, in Lodge, J. (ed.) *The European Community and the Challenge of the Future*, Pinter, London, pp. 89–111.

Shaw, D., Nadin, V. and Westlake, T. (1995) The Compendium of European Spatial Planning Systems, *European Planning Studies*, Vol. 3, no. 3, pp. 390–5.

Shetter, W.Z. (1993) The roots and context of Dutch planning, in Faludi, A. (ed.) *Dutch Strategic Planning in International Perspective*, SISWO, Amsterdam, pp. 67–78.

Sinz, M. (1993) Raumordnung in Europa: Zwischen Geodesign und Politikformulierung, *Raumplanung*, Vol. 60, März, pp. 19–27.

Stanners, D. and Bourdeau, P. (1995) *Europe's Environment: The Dobris Assessment*, European Environment Agency, OOPEC, Luxembourg.

Stoker, G. and Young, S. (1993) Networks: the new driving force, Ch. 8, in *Cities in the 1990s; local choice for a balanced strategy* (Stoker, G. and Young, S.) Longmans, Harlow, pp. 179–90.

Tsoukalis, L. (1983) *The European Community: Past, Present and Future*, Blackwell, Oxford.

Van Breugel, L., Williams, R.H. and Wood, B. (1993) *The Multilingual Dictionary of Real Estate*, E. & F.N. Spon, London.

van der Boel, S. (1994) The challenge to develop a border region. German–Polish cooperation, *European Spatial Research and Policy*, Vol. 1, no. 1, pp. 57–72.

Venturi, M. (1990) *Town Planning Glossary*, K.G. Saur Verlag, München.

Vickerman, R.W. (1992) *The single European market: prospects for economic integration*, Harvester Wheatsheaf, New York.

Vickerman, R.W. (1994) Transport Infrastructure and Region Building in the European Community, *Journal of Common Market Studies*, Vol. 32, no. 1, pp. 1–24.

Williams, R.H. (ed.) (1984) *Planning in Europe. Urban and Regional Planning in the EEC*, Allen & Unwin, London.

Williams, R.H. (1986) The EC Environment Policy, Land Use Planning and Pollution Control, *Policy and Politics*, Vol. 14, no. 1, pp. 93–106.

Williams, R.H. (1988) The European Communities Directive on Environmental Impact Assessment, Chapter 4, in *The role of Environmental Impact in the Planning Process*, Clark, M. and Herington, J. (eds), Mansell, London, pp. 74–87.

Williams, R.H. (1989) Are we speaking the same language? The vocabulary of planning and planning education in languages close to English, in *Proceedings, International Planning Education Conference*, Birmingham Polytechnic, Birmingham.

Williams, R.H. (1990) Supranational environmental policy and pollution control, in Pinder, D. (ed.) *Western Europe: Challenge and Change*, Belhaven, London, pp. 195–207.

Williams, R.H. (1991) Placing Britain in Europe: four issues in spatial planning, *Town Planning Review*, Vol. 62, no. 3, pp. 331–40.

Williams, R.H. (1993a) The Maastricht Treaty: a new status for spatial planning, *Raumplanung*, Vol. 60, März, pp. 6–12.

Williams, R.H. (1993b) Spatial planning for an integrated Europe, in Lodge, J. (ed.) *The EC and the Challenge of the Future*, Pinter, London, pp. 348–59.

Williams, R.H. (1993c) *Blue Bananas, Grapes and Golden Triangles: Spatial Planning for an Integrated Europe. Working Paper* 18, Department of Town and Country Planning, University of Newcastle upon Tyne, Newcastle upon Tyne.

Williams, R.H. (1993d) Baltic gateways, networks and European spatial planning. Paper presented to the 7th Annual AESOP Congress, Lodz, July.

Williams, R.H. (1995a) Contaminated land – a problem for Europe? *Brachflächen-Recycling* 1/95, pp. 19–25.

Williams, R.H. (1995b) European spatial strategies and local development in central Europe, *European Spatial Research and Policy*, Vol. 2, no. 1, pp. 49–61.

Williams, R.H. (1995c) *The European Union Committee of the Regions, its UK Membership and Spatial Planning. Working Paper* 52, Department of Town and Country Planning, University of Newcastle upon Tyne, Newcastle upon Tyne.

Williams, R.H. and Wood, B. (1994) *Urban Land and Property Markets in the UK*, UCL Press, London.

Wood, B. and Williams, R.H. (eds) (1992) *Industrial Property Markets in Western Europe*, E. & F.N. Spon, London.

Wood, C. (1988) The genesis and implementation of environmental impact assessment in Europe, in Clark, M. and Herington, J. (eds) *The Role of Environmental Impact Assessment in the Planning Process*, Mansell, London, pp. 88–102.

Wurzel, R. (1993) Environmental policy, in Lodge, J. (ed.) *The EC and the Challenge of the Future*, Pinter, London, pp. 178–99.

Yuill, D., Allen, K., Bachtler, J. and Wishlade, F. (1995) 15th edn, *European Regional Incentives*, Bowker-Saur, London, and University of Strathclyde, Glasgow.

Zonneveld, W. and D'hondt, F. (1994a) *Aménagement du territoire dans le Benelux*, Commission spéciale pour l'aménagement du territoire, Benelux Economic Union, Brussels.

Zonneveld, W. and D'hondt, F. (eds) (1994b) *Europese ruimtelijke ordening: Impressies en visies vanuit Vlaanderen en Nederland*, NIROV, Den Haag, and VFP, Gent.

Zonneveld, W. and Evers, F. (1995) (eds) *Europa op de plankaart*, NIROV-Europlan, Den Haag.

INDEX